Performance Modeling, Stochastic Networks, and Statistical Multiplexing

second edition

Synthesis Lectures on Communication Networks

Editor
Jean Walrand, *University of California, Berkeley*

Synthesis Lectures on Communication Networks is an ongoing series of 50- to 100-page publications on topics on the design, implementation, and management of communication networks. Each lecture is a self-contained presentation of one topic by a leading expert. The topics range from algorithms to hardware implementations and cover a broad spectrum of issues from security to multiple-access protocols. The series addresses technologies from sensor networks to reconfigurable optical networks.

The series is designed to:

- Provide the best available presentations of important aspects of communication networks.

- Help engineers and advanced students keep up with recent developments in a rapidly evolving technology.

Stochastic Network Optimization with Application to Communication and Queueing Systems
Michael J. Neely
2010

Scheduling and Congestion Control for Wireless and Processing Networks
Libin Jiang and Jean Walrand
2010

Performance Modeling of Communication Networks with Markov Chains
Jeonghoon Mo
2010

Communication Networks: A Concise Introduction
Jean Walrand and Shyam Parekh
2010

Path Problems in Networks
John S. Baras and George Theodorakopoulos
2010

Performance Modeling, Loss Networks, and Statistical Multiplexing
Ravi R. Mazumdar
2009

Network Simulation
Richard M. Fujimoto, Kalyan S. Perumalla, and George F. Riley
2006

Performance Modeling, Stochastic Networks, and Statistical Multiplexing, second edition
Ravi R. Mazumdar

ISBN: 978-3-031-79259-5   paperback
ISBN: 978-3-031-79260-1   ebook

DOI 10.1007/978-3-031-79260-1

A Publication in the Springer series
SYNTHESIS LECTURES ON COMMUNICATION NETWORKS

Lecture #12
Series Editor: Jean Walrand, *University of California, Berkeley*
Series ISSN
Synthesis Lectures on Communication Networks
Print 1935-4185   Electronic 1935-4193

Performance Modeling, Stochastic Networks, and Statistical Multiplexing

second edition

Ravi R. Mazumdar
University of Waterloo, Canada

SYNTHESIS LECTURES ON COMMUNICATION NETWORKS #12

ABSTRACT

This monograph presents a concise mathematical approach for modeling and analyzing the performance of communication networks with the aim of introducing an appropriate mathematical framework for modeling and analysis as well as understanding the phenomenon of statistical multiplexing. The models, techniques, and results presented form the core of traffic engineering methods used to design, control and allocate resources in communication networks. The novelty of the monograph is the fresh approach and insights provided by a sample-path methodology for queueing models that highlights the important ideas of Palm distributions associated with traffic models and their role in computing performance measures. The monograph also covers stochastic network theory including Markovian networks. Recent results on network utility optimization and connections to stochastic insensitivity are discussed. Also presented are ideas of large buffer, and many sources asymptotics that play an important role in understanding statistical multiplexing. In particular, the important concept of effective bandwidths as mappings from queueing level phenomena to loss network models is clearly presented along with a detailed discussion of accurate approximations for large networks.

KEYWORDS

communication networks, performance modeling, point process, fluid inputs, queues, Palm distributions, stochastic networks, insensitivity, effective bandwiths, statistical multiplexing *à ma famille*

Contents

Preface

This monograph is a self-contained introduction to traffic modeling, stochastic networks, and the issue of statistical multiplexing in communication networks. It is a revised and expanded version of *Performance Modeling, Loss Networks, and Statistical Multiplexing*.

Broadly speaking this is the area of performance evaluation of networks. The issue of network performance is one of the key underpinnings of design, resource allocation, and the management of networks. This is because the user experience as well as quality of communication depends on the proper allocation of resources to handle random fluctuations in traffic. From the perspective of a network operator, the installation of network resources or capacity is one of cost management. Since resources (switches, routers, optical fiber links, and wireless base stations) are expensive and hence are limited it leads us to address the issue of efficient network design and consequently the issue of performance. The important characteristics of modern networks are their *largeness* in terms of the number of users, the switch speeds or bandwidths, and the heterogeneity of the applications to be supported. These along with the fact that all users do not connect to the network at the same time in any coordinated way actually helps us to design networks that are remarkably robust and which can provide adequate user experience most of the time. At the base of all of this is the issue of *statistical independence* and this mathematical phenomenon allows us to design networks based on *average* behavior rather than on *worst-case* behavior. Of course, in an adverse case the network performance does go down as for example when disasters take place whereby simultaneous correlated behavior of users tends to overwhelm network resources leading to degraded or very poor performance. However, designing networks for such rare scenarios is simply too expensive.

The purpose of this monograph is to introduce the basics of traffic modeling and develop the appropriate mathematical framework for understanding how the statistical behavior of traffic impacts the performance. Often, resulting "smoothness" manifests in the fact that knowledge of *average* behavior is all that is needed to quantify several important performance characteristics. The first is the notion of *insensitivity* in stochastic networks. The second is the notion of *statistical multiplexing* where-by many more users or sources can be handled by a system than would be expected based on their so-called peak requirements provided we allow some leeway in performance constraints. The objective is to introduce the appropriate level of mathematical modeling and analysis so as to be able to obtain useful models and to analyze the effects interaction in networks. At the same time, the aim is to introduce the reader to various results from stochastic analysis that allow us a more detailed view of the underlying models. The monograph is not exhaustive in the coverage of the topics but a detailed bibliography with some background is provided at the end of each chapter. However a sufficient amount of detail is given to have pedagogic value so that the reader develops some understanding of the basic modeling and tools of performance evaluation. A key departure from the number of excellent books on the subject is the emphasis on the sample-path or dynamic behavior of the underlying processes. The background that is assumed is a basic course in random processes and especially an understanding of the laws of large numbers, the Central Limit Theorem, and a basic understanding of Markov chains and ergodicity. These are recalled in the Appendix. The development starts from elementary properties of the underlying stochastic models leading up to issues and concepts that form the crux of performance evaluation. Wherever possible and to aid understanding, proofs of the results have been included. Every attempt has been made to include mathematical rigor while keeping in mind the audience- graduate students who wish to acquire modeling skills as well as analytical tools. The approach that has been adopted is to derive the key queueing and performance formulae by exploiting the basic structures of the underlying mathematical models namely the stationarity and evolution of the sample-paths.

Chapter 1 introduces the basic probabilistic models of traffic in networks. The most basic model is that of the Poisson process and yet the structure is so rich that a good understanding is necessary to be able to appreciate some of the future results. From Poisson processes we then introduce the more general class of stationary point processes and we will see some key concepts that are necessary for performance evaluation, namely the inspection paradox and forward recurrence times. We will then see some more general traffic models that are necessitated by the emergence of new applications and traffic patterns in modern networks. The chapter ends with an introduction to the notion of Palm distributions (or conditioning at points) and the important notion of PASTA (Poisson Arrivals See Time Averages) or EATA (Event and Time averages). These ideas are not simply mathematical artifacts, they constitute a measurement reference that is the basis of performance received by users.

Chapter 2 introduces queueing models. Here the intention is to take a very specific view at queueing models- one from the view point of performance-rather than any systematic or exhaustive view presentation of queueing models. Three key results are of interest here: 1) The idea of the Lindley equation, 2) The so-called Pollaczek-Khinchine formula for mean waiting times and 3) Little's formula. These can be obtained under fairly general hypotheses and this is the approach followed. One of the key ideas that we introduce is that all the well known queueing formulae are simple consequences of stationarity and the sample-path evolution. This is best captured by the so-called Rate Conservation Law (RCL) that we introduce and use extensively to obtain Little's Law and the Pollaczek-Khinchine formula. One of the advantages of this approach is that the explicit dependence on different types of underlying measures is brought to the forefront. We also discuss an extension of the RCL termed the Swiss Army formula. The chapter concludes with the formulae for fluid queueing models that are of increasing importance in today's networks.

Chapter 3 deals with loss models. The chapter begins with treatment of the infinite server model, especially the so-called $M/G/\infty$ model because of the importance that it plays in the context of loss models that arise in the context of statistical multiplexing. In particular, we discuss the issue of insensitivity of such models and study the output process because of its importance in the network context. We then study the classical Erlang and Engset models where the idea of Palm distributions is clearly seen. We then study the generalization of the Erlang model known as the multirate Erlang model that is of relevance in today's networking context where the traffic granularity is in terms of bits rather than discrete packets because of the speeds involved. In particular, we focus on the computation of blocking probabilities. We do this via the so-called Kaufman-Roberts algorithm and an analytical approach that is possible when systems are large. The chapter concludes with the loss network model. In this context the chapter covers the basic ideas of the Erlang fixed point approximation, its accuracy and concludes with the large system case where we can exploit ideas from large-deviations , local limit theorems to obtain explicit results..

Chapter 4 introduces queueing network or stochastic network models. It begins with a discussion of classical Jackson network models that require Markovian hypotheses on the inputs and service requirements. We then introduce the notion of Whittle networks that allows for the relaxation on the requirement that services are exponentially distributed and results in insensitive stationary distributions. There is a strong connection between such models and so-called flow models that are useful for today's networks and we discuss some recent results in this direction. In particular the relationship between insensitive bandwidth allocation strategies and network utility optimization models is presented. The notion of *chaoticity* in networks is then introduced whereby statistical independence emerges amongst a set of interacting nodes when a network is large and in particular this is used to analyze a randomized load balancing scheme called Join the Shortest Queue mechanism. The chapter concludes with a brief discussion of fluid networks.

Chapter 5 provides a basis for studying statistical multiplexing from an analytical viewpoint. We first introduce the multiplexing in the context of mean delay constraints. Via a use of the Pollaczek-Khinchine formula we show the role of the loss model in this context and introduce an important concept of *effective bandwidths*. This idea has been one of the important conceptual advances in the last 15 years not only in the context of networking but also in trying to quantify the notion of *burstiness*. From mean delay constraints we move on to packet loss constraints that emerged as important performance criteria that drove the appellation of the acronym QoS (Quality of Service). In both these contexts we see that the effective bandwidth is the right level of abstraction to capture the idea of statistical multiplexing and also allows us to go from the microscopic level of analyzing queueing behavior to the macroscopic level of loss models.

An Appendix at the end collects some basic mathematical background on the main probabilistic tools used.

These notes are suitable for teaching a topics course on modeling and performance evaluation as a part of a larger course in networking. Some of the material in the last two chapters, particularly Sections 3.5, 3.6, and 3.7, as well as Sections 5.5 and 5.6 can be omitted in a first pass, depending on the student background and interest.

This monograph has benefited from interactions with many individuals over the years. I would like to especially thank my friends and collaborators Pierre Brémaud, Fabrice Guillemin, and Nikolay Likhanov for teaching me all that I know about the mathematical tools and frameworks for performance evaluation. Chapters 3, 4, and 5 have been inspired by the work of Frank Kelly who I wish to thank for his comments. I have benefited from their insights, deep mathematical culture and I am richer for their friendship. Various versions of these notes have been presented to unsuspecting students here at Waterloo and at Purdue University, and I wish to thank them for their feedback and pointing out of the errors. I am indebted to my colleague Patrick Mitran for going through the notes and pointing out many typos and errors. Any errors that remain are due to my negligence alone.

The author wishes to thank Jean Walrand of the University of Berkeley for inviting him to submit this monograph to the series on Synthesis Lectures in Networking. The author also wishes to thank the publisher, Michael Morgan, for the editorial assistance in the production.

I owe a debt of thanks to my family: my wife Catherine Rosenberg and my children Claire and Eric, for being there and for their encouragement, love and inspiration; and for making the world a cheerful place for to be in. Together with my larger family they form my *raison d'etre* and I am forever grateful for their support.

Ravi R. Mazumdar

Waterloo, November 2009
Revised May 2013

CHAPTER 1

Introduction to Traffic Models and Analysis

1.1   INTRODUCTION

To understand the need for traffic modeling let us begin by looking at how the current Internet evolved and the expectations of users and operators of networks. We start by looking at the Internet as it is today. Loosely speaking, the communications infrastructure is just a mechanism to transfer information that is encoded into bits and packaged into packets across distances on a wired or wireless medium, i.e., a *bit pipe*. The architecture of the internet is based on a *best effort* paradigm. By this it is meant that the network does not offer any performance guarantees. All the network aims is to provide a reliable and robust packet transfer mechanism that is based on TCP/IP. The salient features of packet transport on the Internet are:

- No performance guarantees i.e., QoS (Quality of Service), or admission control is provided for explicitly. Best effort only.

- TCP (Transmission Control Protocol) is an end-to-end protocol residing on hosts that provides

  - Reliable and sequential packet transport over the IP's (Internet Protocol) unreliable and non-sequential service.

  - Sender-receiver flow control that adjusts flow rate according to a bottleneck along a path.

- TCP performs congestion control (rate at which packets are injected) based on the use of an adaptive window that reacts to packet loss or unacknowledged packets.

- The network also offers UDP (unreliable datagram protocol) over IP that does not guarantee loss free delivery. This form is usually not liked by network operators due to lack of reactivity of flows to congestion in the network.

As applications have become more numerous and there is a greater share of so-called real-time services being offered, users have become more demanding of the performance afforded to them by network providers. This has brought into focus the need for providing *Quality of Service* (QoS) guarantees usually specified by delays and packet drop or loss rates. As the demands of applications keep on escalating, network operators are finding that to cope and provide QoS guarantees there is a need to plan and better manage network resources. One way to manage resources over the current Internet is to provide some sort of prioritized treatment of different types of requests, i.e., treat packets of different types of applications differently by giving priority to real-time over purely data traffic, blocking packets from certain types of applications during peak hours, etc. Even then, the performance guarantees are only *best effort* since the underlying transport mechanism is based on TCP. In order to provide better guarantees and also to utilize resources better there is a clear need to understand the effect of traffic characteristics on network resources. Given the trend is towards providing QoS, network operators must devote resources to handling different types of requests, especially if they are going to charge according to the service provided. In order to do so network operators must be able to size or dimension the required network resources (how much server capacity (speed), numbers of servers, buffering requirements, routing topology, etc) to handle the demands, prevent the network from becoming overloaded or prevent traffic that is not exigent from consuming valuable resources. For this we need a good understanding of traffic or packet flows. This is the purview of *traffic engineering*.

This book is motivated by the need to provide an understanding of tools and frameworks to perform traffic engineering. However, this is not the way the current internet has evolved. This is partly because of a mistaken belief that modeling and analysis have a limited role to play. However, as will be shown, it is precisely the development of mathematical analysis that has led to *rules of thumb* and yet there are situations where our intuition might not be very useful. In this book we will see the basic building blocks that allow us to develop reasonable insights and rules that allow for the efficient allocation of resources in a network, and understand the influences of various parameters and strategies on network performance.

1.1.1   QUANTITATIVE TOOLS

To design a network we need both qualitative and quantitative results about:

- Capacities or speed of transmission links (referred to as bandwidth).

- The time-scales of interest.

- The number of transmission links required.

- Rules for aggregation and dis-aggregation of traffic.

- Buffering requirements.

- Size of switches (number of ports, types of interconnections, input or output buffering, etc.)

- How packets or bits are served and the impact of such choices on resulting mathematical model.

Because it is impossible to exactly predict human behavior and their communication require-ments (type of communication and devices used) one needs a set of mathematical tools drawn from stochastic processes. Moreover, we also need parsimonious models because we would like to obtain insights and thus having as few parameters as possible to characterize the statistical characteristics is desirable. In order to provide predictability, we would also like to know whether the statistics change or not.

It turns out that using the tools of probability and stochastic processes, queueing theory in particular, gives us very a powerful framework for predicting network behavior and thus leads to good rules for designing them. One of the key assumptions will be that of stationarity (in a statistical sense). Although, stationarity is very special, it turns out that one indeed can identify periods when the network performance can be well predicted using stationary models. Thus, this will be an implicit assumption in the sequel unless specifically mentioned to the contrary.

Network performance is usually defined in terms of statistical quantities called the QoS parameters.

They are:

- Call blocking probabilities in the context of flows and circuit-switched architectures–often called the Grade of Service (GoS)

- Packet or bit loss probabilities

- Moments and distributions of packet delay, denoted D, such as the mean delay $E[D]$, the variance $var(D)$ which is related to the *jitter*, the tail distribution $\mathbb{P}(D \geq t)$, etc.

- The mean or average throughput (average number of packets or bits transmitted per sec), usually in kbits/sec, Mbits/sec, etc.

- The transient and dynamic behavior of networks like the duration of overloads or congestion, etc.

The sources of randomness in the internet arise from:

- Call, session, and packet arrivals are unpredictable (random).

- Holding times, durations of calls, file sizes, etc are random.

- Transmission facilities, switches, etc are shared.

- Numbers of users are usually not known a priori, and they arrive and depart randomly.

- Statistical variation, noise, and interference on wireless channels.

Queueing arises because instantaneous speeds exceed server or link capacities momentarily because bit flows or packet flows depend on the devices or routers feeding a link. Queueing leads to delays and these delays depend not only on traffic characteristics but the way packets are processed. The goal of performance analysis is to estimate the statistical effects of variations. In the following we begin by first presenting several useful models of traffic, in particular ways of describing packet or bit arrivals.

We first begin by introducing models for traffic arrivals.

1.2 TRAFFIC ARRIVAL MODELS

It must be kept in mind that the primary purpose of modeling is to obtain insights and qualitative information. There are some situations when the models match empirical measurements. In that case, the analytical results also provide quantitative estimates of performance. To paraphrase the words of the famous statistician Box, *models are always wrong but the insights they provide can be useful*. It is with this caveat that one must approach the issue of traffic engineering.

As far as studying the performance of networks, the key is to understand how traffic arrives and how packets or bits are processed. Modeling these processes is the basic building blocks of queueing theory. However, even in this case there is no one way to do it. It depends on the effect we are studying and the time scale of relevance.

Let us look at this issue a bit further. Packet arrivals can be as discrete events when packets arrive on a link or router buffer, or in the case when a source transmits at very high speed and viewed at the time-scale of bits one could think of arrivals as fluid with periods of activity punctuated with periods of inactivity. The former leads to so-called *point process* models of arrivals while the latter are called *fluid* inputs. In essence we are just viewing arrivals at different time-scales. Both models are relevant in modeling real systems but we will restrict our selves to point process models for the most part. The two arrival patterns are illustrated below.

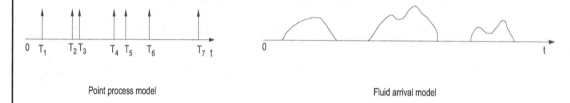

Point process model   Fluid arrival model

Figure 1.1: Discrete and fluid traffic arrival models.

Definition 1.1 A stochastic process is called a simple point process if it is characterized as follows:
Let $\{T_n(\omega)\}_{n=-\infty}^{\infty}$ be a real sequence of random variables (or points) in $\mathbb{R}$ with $\cdots < T_{-1}(\omega) < T_0 \leq 0 < T_1(\omega) < T_2(\omega) < \cdots$ and $\lim_{n \to \infty} T_n(\omega) = \infty$ *a.s.*.
Define N_t as :

$$N_t(\omega) = \sum_n \mathbb{1}_{(0 < T_n(\omega) \leq t)} \tag{1.1}$$

Then N_t is the counting process of $\{T_n\}$ and is commonly called a point process.

Remark 1.2

1. N_t counts the number of points T_n that occur in the interval $(0, t]$. The notation $N(a, b]$ is also used to indicate the number of points in $(a, b]$. Thus, $N_t = N(0, t]$.

2. $\lim_{t \to \infty} N_t = \infty$ $a.s.$ by definition.

3. The random variables $S_n = T_n - T_{n-1}$ are called the inter-arrival times.

4. The event $\{N_t = n\} = \{T_n \le t < T_{n+1}\}$.

5. If $\{S_n\}$ are identically distributed then N_t is called a stationary simple point process. If $\{S_n\}$ are i.i.d then N_t is called a *renewal process*. In particular, if $\{S_n\}$ are i.i.d. and exponentially distributed then N_t is called a Poisson process .

6. By definition of a simple point process, two events or arrivals cannot take place simultaneously (since $T_n < T_{n+1} < T_{n+2} \cdots$).

The Poisson process plays very an important role in traffic theory. Before introducing it and studying its properties let us look a bit ahead to the type of results we will need.

The first class of results we will need are results on superposition or multiplexing of traffic streams that amount to aggregation of point processes. The second type of result we need are results related to de-multiplexing or dis-aggregation when traffic streams split as shown below.

Multiplexing   De-multiplexing

Figure 1.2: Typical traffic mixing scenarios of interest.

But before doing so, we state a key result from probability we use frequently.

Theorem 1.3   Strong Law of Large Numbers (SLLN)

Let $\{X_i\}$ be a collection of i.i.d. r.v's with $\mathbb{E}|X_0| < \infty$ and $\mathbb{E}[X_0] = m < \infty$. Then

$$\lim_{N \to \infty} \frac{1}{N} \sum_{k=1}^{N} X_k = m \quad a.s. \tag{1.2}$$

This is called an ergodic theorem and it is related to but slightly different from the notion of an ergodic process which implies that

$$\lim_{N \to \infty} \frac{1}{N} \sum_{k=1}^{N} f(X_k) = \mathbb{E}[f(X_0)] \quad a.s .$$

for all *nice* (measurable) functions $f(.)$. For a process $\{X_n\}$ to be ergodic it is not necessary for $\{X_n\}$ to be i.i.d. or even a stationary sequence.

Note the SLLN and the ergodic theorem can be extended to continuous time stochastic processes $\{X_t\}_{t\geq 0}$ in the obvious way, i.e.,

$$\lim_{T\to\infty} \frac{1}{T} \int_0^T f(X_s)ds = \mathbb{E}[f(X_0)]\ \ a.s.$$

Let us see a basic property of renewal processes. It has to do with the existence of an average rate. This is stated and shown below.

Proposition 1.4

Let $\{S_n\}$ denote the inter-arrival times of a renewal process N_t and let $\mathbb{E}[S_0] = \frac{1}{\lambda_N}$ where $\lambda_N > 0$. Then

$$\boxed{\lim_{t\to\infty} \frac{N_t}{t} = \lambda_N\ \ a.s.} \tag{1.3}$$

and λ_N is called the average rate or average intensity of the point process.

Proof. First note, by the definition of N_t we have $T_{N_t} \leq t < T_{N_t+1}$ from the remark above. Moreover, $T_{N_t} = \sum_{k=1}^{N_t} S_k$ by the definition of S_k. Therefore,

$$\frac{1}{N_t} \sum_{k=1}^{N_t} S_k \leq \frac{t}{N_t} \leq \frac{1}{N_t} \sum_{k=1}^{N_t+1} S_k$$

Since $\{S_i\}$ are i.i.d. and of finite mean, by the SLLN both the leftmost and rightmost terms go to $\mathbb{E}[S_0] = \frac{1}{\lambda_N}$ as $t \to \infty$ (since $N_t \to \infty$) and hence the result follows.   □

Definition 1.5   Poisson Processes
A renewal process $\{N_t\}_{t\geq 0}$ is said to be a Poisson process with rate or intensity λ if the inter-arrival times $\{S_n\}$ are i.i.d. exponential with mean $\frac{1}{\lambda}$.

An immediate consequence (and often used as the definition without the need for the introduction of inter-arrival times) for Poisson processes is the following that is more commonly found in texts:

Proposition 1.6   Equivalent Definition of Poisson Processes

Let $\{N_t\}_{t\geq 0}$ be a Poisson process with intensity λ. Then

1. $N_0 = 0$

2. $\mathbb{P}(N_t - N_s = k) = \frac{(\lambda(t-s))^k}{k!} e^{-\lambda(t-s)}$ *for* $t > s$.

3. $\{N_{t+s} - N_s\}$ *is independent of* N_u, $u \leq s$.

An immediate consequence of the above definitions is that:

$$\begin{aligned}\mathbb{P}(N_{t+dt} - N_t = 1) &= \lambda dt + o(dt) \\ \mathbb{P}(N_{t+dt} - N_t \geq 2) &= o(dt)\end{aligned}$$

Note the last property states that there cannot be more than one jump taking place in an infinitesimally small interval of time or equivalently two or more arrivals cannot take place simultaneously.

As mentioned above, the above proposition is often used to define Poisson processes. Indeed, the proof relies on some crucial properties of the exponential distribution and so let us study some of them.

Recall, a nonnegative r.v. X is said to be exponentially distributed with parameter λ, denoted as $exp(\lambda)$ if $\mathbb{P}(X > t) = e^{-\lambda t}$ where $\lambda > 0$. Hence, its probability density function $p(t)$ is just $\lambda e^{-\lambda t}$, $t > 0$. The following quantities associated with exponentially distributed r.v's are used often.
If X is $exp(\lambda)$ then $\mathbb{E}[X] = \frac{1}{\lambda}$ and $\mathbb{E}[X^2] = \frac{2}{\lambda^2}$.
However, there are more important properties that are unique to exponential distributions.

Proposition 1.7 Memoryless Property

Let X *be* $exp(\lambda)$. *Then for any* $t, s > 0$:

$$\mathbb{P}(X > t + s | X > s) = \mathbb{P}(X > t) = e^{-\lambda t}$$

Moreover, the property remains true if we replace s by a random variable $S(\omega)$ *independent of* X *i.e.:*

$$\boxed{\mathbb{P}(X > t + S(\omega) | X > S(\omega)) = \mathbb{P}(X > t) = e^{-\lambda t}}$$ (1.4)

Proof. Since X is exponential:

$$\begin{aligned}\mathbb{P}(X > t + s | X > s) &= \frac{\mathbb{P}(X > t + s, X > s)}{\mathbb{P}(X > s)} = \frac{\mathbb{P}(X > t + s)}{\mathbb{P}(X > s)} \\ &= \frac{e^{-\lambda(t+s)}}{e^{-\lambda s}} = e^{-\lambda t}\end{aligned}$$

The proof of the second result follows by conditioning on $S(\omega)$ and using independence. □

Remark 1.8

The memoryless property states that if the r.v. is exponentially distributed, it does not matter how long we have waited, the distribution of the remaining wait time is the same going forward (it forgets how long we have already waited). An interesting consequence of this property is, for any $a > 0$:

$$\begin{aligned} \mathbb{E}[X|X > a] &= a + E[X] \\ var(X|X > a) &= var(X). \end{aligned}$$

The following property is also useful to remember. Let $\{X_i\}_{i=1}^n$ be a collection of n independent $exp(\lambda_i)$ exponential r.v's. Then the r.v. $Z = \min\{X_1, X_2, \ldots, X_n\}$ is exponentially distributed with parameter $\lambda_1 + \lambda_2 + \cdots + \lambda_n$. This just follows from the fact that the complementary distribution function $1 - F(t)$ where $F(t) = \mathbb{P}(Z \leq t)$ is the distribution function of the minimum of n independent r.v.s $\{X_1, X_2, \ldots, X_n\}$ and is given by $1 - F(t) = \prod_{i=1}^n (1 - F_i(t))$ where $F_i(t)$ is the distribution function of X_i. Moreover, Z is independent of the r.v. that is the minimum i.e., $\mathbb{P}(Z \leq t | X_j = \min\{X_1, \cdots, X_n\}) = \mathbb{P}(Z \leq t) \; \forall j = 1, 2, \cdots, n$.

Now taking as convention the process starting at 0 i.e., $T_0 = 0$,

$$T_n = \sum_{i=1}^n (T_i - T_{i-1})$$

The distribution of the n-th jump time or point of a Poisson process,i.e. T_n, is called the *Erlang-n distribution* or an Erlang distribution with n stages. We derive it below.

As we have seen N_t is just the process that counts how many jumps have taken place before t, moreover the following two events are equivalent: $\{N_t = n\} = \{T_n \leq t < T_{n+1}\}$. Since $T_i - T_{i-1}$ are i.i.d. exponential mean $\frac{1}{\lambda}$,

$$\begin{aligned} \mathbb{P}(T_n \leq t) &= \mathbb{P}(N_t \geq n) \\ &= \sum_{k=n}^\infty \mathbb{P}(N_t = k) \\ &= \sum_{k=n}^\infty \frac{(\lambda t)^k}{k!} e^{-\lambda t} \end{aligned}$$

Denoting the density of T_n by $p_{T_n}(t)$ we have :

$$p_{T_n}(t) = \frac{d}{dt}\mathbb{P}(T_n \leq t) = \lambda \frac{(\lambda t)^{n-1}}{(n-1)!} e^{-\lambda t} \tag{1.5}$$

Thus, an Erlang-1 r.v. is an exponential r.v. and Erlang-k is the sum of k i.i.d exponential random variables. Noting that the characteristic function of an exponential r.v. is given by

$$C(h) = \frac{\lambda}{\lambda - ih} \; .$$

If the mean of the Erlang-k r.v. is kept fixed at $\frac{1}{\lambda}$ as the number of stages increases i.e., $\lambda_k = \frac{\lambda}{k}$ is the rate of each exponential, then:

$$\lim_{k \to \infty} \left(\frac{\lambda_k}{\lambda_k - ih} \right)^k = e^{ih\frac{1}{\lambda}}$$

that is the characteristic function of a deterministic r.v. with value $\frac{1}{\lambda}$ i.e., the Erlang-k r.v. converges to a deterministic r.v. of length $\frac{1}{\lambda}$. In other words, keeping the mean constant and increasing the number of stages of an Erlang-k r.v. makes it more and more deterministic!

Here, are some more properties of interest about Poisson processes which are stated without proof.

Proposition 1.9 *Let $\{N_t\}_{t \geq 0}$ be a homogeneous Poisson process with rate or intensity $\lambda > 0$.*

1. Mean $\mathbb{E}[N_t] = \lambda t$

2. Variance $\mathbb{E}[(N_t)^2] - [\mathbb{E}[N_t]]^2 = \lambda t$

3. Characteristic Function $C(h) = \mathbb{E}[e^{ihN_t}] = e^{\lambda t (e^{ih} - 1)}$

4. Let $\pi_n(\lambda t) = \mathbb{P}(N_t = n)$ denote the Poisson distribution with parameter λt. Then:

$$\mathbb{P}(N_t > n) = t \int_0^\lambda \pi_n(\mu t) d\mu$$

5. No common points If N_t^1 and N_t^2 are two independent Poisson processes with intensities λ_1 and λ_2 respectively, then they cannot have common jumps or points i.e., $\mathbb{P}(\Delta N_t^1 = \Delta N_t^2 = 1) = 0$ where $\Delta N_t = N_t - N_{t-}$ denotes a jump at time t..

We now see the first set of results that yield rules for traffic aggregation and dis-aggregation.

Proposition 1.10 Superposition and thinning of Poisson processes

1. Let N_t^1 and N_t^2 be two independent Poisson processes with intensities λ_1 and λ_2 respectively. Then the process $N_t = N_t^1 + N_t^2$ is Poisson with intensity $\lambda_1 + \lambda_2$.

2. Suppose we randomly select points $\{T_n\}$ of a Poisson process of intensity λ with probability p and define a new point process $\hat{N}_t$ as the resulting point process composed of the selected points. Then $\{\hat{N}_t\}$ is Poisson with intensity λp.

Proof. 1) There are many ways of showing this result: here are 3 ways based on the different characterizations of the Poisson process given above.

Method 1: Suppose $N_t = N_t^1 + N_t^2$. The points of N_t are the points of N^1 and the points of N^2 denoted by τ_i are defined as follows:

$$
\begin{aligned}
\tau_1 &= \min\{T_1^1, T_1^2\} \\
\tau_2 &= \min\{T_j^i > \tau_1\} \\
\tau_i &= \min\{T_j^k > \tau_{i-1}\}
\end{aligned}
$$

In other words, we pick the points in the order they occur noting that they could be either belonging to N^1 or to N^2. Note because of independence, the probability that any two points coincide is 0.

Now we use the fact that for the first point, it is the min of two independent exponentially distributed r.v.'s, and this is exponential with parameter $\lambda_1 + \lambda_2$. Now because exponential r.v.'s have a memoryless property, going forward it is also the first occurrence of a point for two independent exponential r.v.'s and has the same distribution as the first point, and so on. Thus, the inter-arrival times of N_t are exponential with parameter $\lambda_1 + \lambda_2$ or N_t is Poisson $\lambda_1 + \lambda_2$.

Method 2: This is by direct calculation

$$
\begin{aligned}
\mathbb{P}(N_t = k) &= \sum_{j=0}^{k} \mathbb{P}(N_t^1 = j, N_t^2 = k - j) \\
&= \frac{1}{k!} \sum_{j=0}^{k} k! \frac{(\lambda_1 t)^j}{j!} \frac{(\lambda_2 t)^{k-j}}{(k-j)!} e^{-(\lambda_1 + \lambda_2)t} \\
&= \frac{((\lambda_1 + \lambda_2)t)^k}{k!} e^{-(\lambda_1 + \lambda_2)t}
\end{aligned}
$$

where we have used the binomial formula $(x + y)^n = \sum_{k=0}^{n} \binom{n}{k} x^k y^{n-k}$.

Method 3: Use the characteristic function of a Poisson process:

$$
C(h) = \mathbb{E}[e^{ihN_t}] = e^{\lambda t(e^{ih} - 1)}
$$

and $C_N(h) = C_{N^1}(h) C_{N^2}(h)$ because of independence.

2) We shall show a bit more. Let N_t^1 denote the point process if we select the points of N_t (with probability p) and N_t^2 be the point process that consists of the points that are not picked. We will see both N_t^1 and N_t^2 are Poisson processes with intensities λp and $\lambda(1 - p)$, respectively, and moreover are independent of each other.

First note that $\mathbb{P}(N_t^1 = k | N_t = n) = \binom{n}{k} p^k (1-p)^{n-k}$. Then,

$$
\begin{aligned}
\mathbb{P}(N_t^1 = k, N_t^2 = m) &= \mathbb{P}(N_t^1 = k, N_t = k+m) \\
&= \binom{k+m}{k} p^k (1-p)^m \frac{(\lambda t)^{k+m}}{(k+m)!} e^{-\lambda t} \\
&= \frac{(\lambda p t)^k}{k!} e^{-\lambda p t} \frac{((1-p)\lambda t)^m}{m!} e^{-(1-p)\lambda t} \\
&= \mathbb{P}(N_t^1 = k) \mathbb{P}(N_t^2 = m),
\end{aligned}
$$

showing that both N_t^1 and N_t^2 are Poisson with intensities λp and $\lambda(1-p)$, respectively, and are independent. $\qquad\square$

Remark 1.11 There is an important result due to Watanabe: if a stochastic process X_t has stationary independent increments and the sample-paths are piecewise discontinuous then X_t must be a Poisson process.

The superposition of independent Poisson processes and the independent thinning of a Poisson process are two of the basic types of results we need for multiplexing and demultiplexing traffic streams. A natural question is whether such results can be obtained for other types of point processes. For example, if N_t^1 and N_t^2 are independent renewal processes is their sum renewal? The answer is no. However, one key result is that on the superposition of a large number of independent simple point processes whose limit is Poisson. The result is stated without proof but it follows from the convergence to a Poisson random variable of sums of independent Bernoulli r.v.'s with vanishing but summable probabilities.

Proposition 1.12 Superposition of Independent Stationary Point Processes

Let $\{N_t^i\}_{i=1}^N$ be a collection of independent stationary simple point processes with the following properties:

1. $\lim_{N \to \infty} \sum_{i=1}^N \mathbb{E}[N_1^i] = \lambda$

2. $\sum_{i=1}^N \mathbb{P}(N_1^i \geq 2) \to 0$ *as* $N \to \infty$

   Then

$$
\lim_{N \to \infty} \sum_{i=1}^N N_t^i \xrightarrow{\text{weakly}} N_t
$$

for every fixed $t < \infty$ where N_t is a Poisson r.v. with parameter λt.

One way of interpreting this result is that if a very large number of independent point processes of negligible intensity are multiplexed such their sum intensity is finite, then the superposition behaves as a Poisson process. This result justifies the central role that the Poisson process plays in networking because it can be used at the packet level, at the level of session arrivals, etc., in scenarios when a large number of users are present.

1.2.1   NON-HOMOGENEOUS POISSON PROCESS

In many applications the assumption that the point process is stationary might not be valid and we are forced to do away with the assumption that the increments are stationary. Of course in order to develop useful insights and to perform calculations we need to impose some hypotheses. A natural extension to the Poisson process is the non-stationary or non-homogeneous Poisson process defined by the following properties:

Definition 1.13 $\{N_t\}_{t \geq 0}$ is said to be a non-homogeneous Poisson process with intensity λ_t if:

1. $N_0 = 0$

2. For $t > s$, $\mathbb{P}(N_t - N_s = k) = \frac{1}{k!} \left(\int_s^t \lambda_u du \right)^k e^{-\int_s^t \lambda_u du}$

3. $\{N_{t+s} - N_s\}$ is independent of N_u, $u \leq s$.

Of course since λ_t depends on t we need to impose some conditions on the point process for it to be well-behaved, i.e., the points do not accumulate. For this it is necessary and sufficient that $\int_0^\infty \lambda_s ds = \infty$.

With this property it immediately follows that $\{\lim_{t \to \infty} N_t = \infty$ a.s.$\} \iff \lim_{t \to \infty} \int_0^t \lambda_s ds = \infty$ and $0 < T_1 < T_2 < \cdots < T_n < T_{n+1} < \cdots$ and once again:

$$\mathbb{P}(N_{t+dt} - N_t = 1) = \lambda_t dt + o(dt)$$
$$\mathbb{P}(N_{t+dt} - N_t \geq 2) = o(dt)$$

Note the last property states that there cannot be more than one jump taking place in an infinitesimally small interval of time or two or more arrivals cannot take place simultaneously.

We now see an important formula called Campbell's Formula which is a smoothing formula.

Proposition 1.14   *Let $\{N_t\}_{t \geq 0}$ be a non-homogeneous Poisson process with intensity λ_t. Let $\{X_t\}$ be a continuous-time stochastic process such that $N_t - N_s$ is independent of $X_u, u \leq s$ and let $f(.)$ be a continuous and bounded function. Then:*

$$\boxed{\mathbb{E}[\sum_{n=1}^{N_t} f(X_{T_{n-}})] = \mathbb{E}[\int_0^t f(X_s) \lambda_s ds]} \tag{1.6}$$

Proof. It is instructive to see the proof. Let $p_n(t)$ be the probability density of the n-th point T_n (which has a density given by the Erlang-n distribution). Then from (1.5) noting that the Poisson process has the density λ_t,

$$p_n(t) = \lambda_t \frac{(\int_0^t \lambda_s ds)^{n-1}}{(n-1)!} e^{-\int_0^t \lambda_s ds}$$

from which it readily follows that $\sum_{n=0}^{\infty} p_n(t) = \lambda_t$.

Therefore,

$$
\begin{aligned}
\mathbb{E}[\sum_{n=1}^{N_t} f(X_{T_n-})] &= \mathbb{E}\int_0^t [\mathbb{E}[\sum_{n=1}^{\infty} f(X_{s-})|T_n = s]ds] \\
&= \mathbb{E}[\int_0^t \sum_{n=1}^{\infty} f(X_{s-}) p_n(s) ds] = \int_0^t \mathbb{E}[f(X_s)\lambda_s]ds
\end{aligned}
$$

where X_{s-} can be replaced by X_s because of continuity (in s).

Also since λ_t is non-random it can be removed from the expectation. $\square$

Remark 1.15

1. Note $\sum_{n=0}^{N_t} f(X_{T_n})$ can be written as $\int_0^t f(X_s)N(ds)$ where the integral is the so-called Stieltjes integral.

2. In particular, if N_t is homogeneous Poisson with intensity λ and X_t is stationary then:

$$
\mathbb{E}[\int_0^t f(X_s)N(ds)] = \lambda \int_0^t \mathbb{E}[f(X_s)]ds = \lambda t \mathbb{E}[f(X_0)]
$$

   This is the so-called PASTA property that will be discussed a bit later.

3. More generally, a point process N is said to be a Poisson process in $\mathbb{R}^d$ (d-dimensional) with intensity $\lambda(x)$, $x \in \mathbb{R}^d$ if for any $A, B \subset \mathbb{R}^d$ and $A \cap B = \phi$ then $N(A)$ and $N(B)$ are independent, and the distribution of $N(A)$ is given by the Poisson distribution with parameter $\int_A \lambda(x)dx$. Let $\mathbf{0}$ denote the origin. Suppose N is a Poisson process in $\mathfrak{R}^2$ with intensity λ, then the probability that there is no point within a radius r of $\mathbf{0}$ is just $e^{-\pi r^2 \lambda}$ (i.e., the distance to the nearest point is exponentially distributed).

Besides the obvious time dependence, non-homogeneous Poisson processes have another important application in the modeling of networks. Consider the case of a so-called marked Poisson process which is the basic model that is used in the modeling of queues. This is characterized by arrivals which take place as a Poisson process with rate or intensity say λ. Each arrival brings a *mark* associated with it denoted by σ_n which are assumed to be i.i.d. In applications the mark denotes the amount of work or packet length associated with an arriving packet. Then the process:

$$
X_t = \sum_n \sigma_n \mathbb{1}_{(0 < T_n \le t]} \tag{1.7}
$$

is called a Marked Poisson Process if the $\{T_n\}$ correspond to the points of a Poisson process.

Suppose $\{\sigma_n\}$ are i.i.d. with common probability density $\sigma(x)$, then we can think of X_t as a two-dimensional point process whose points are $(T_n, \sigma_n) \in \Re^2$.

Let $A_i \subset \Re^2$ be defined as follows: $A_i = \{(x, y) : x \in (a_i, b_i], \; y \in (c_i, d_i]\}$ and let:

$$X(A_i) = \{\# \; of \; n : (T_n, \sigma_n) \in A_i\}$$

If $\{A_i\}_{i=1}^m$ is collection of disjoint sets in $\Re^2$ it is easy to see from the definition above that $\{X(A_i)\}_{i=1}^m$ are independent and moreover $X(A_i)$ is a Poisson r.v. with

$$\mathbb{P}(X(A_i) = n) = \frac{\mu(A_i)^n}{n!} e^{-\mu(A_i)}$$

where $\mu(A_i) = \lambda |b_i - a_i| \int_{c_i}^{d_i} \sigma(x) dx$ by independence of $\{T_n\}$ and $\{\sigma_n\}$ and so in particular X_t can be seen as a two-dimensional non-homogeneous Poisson process with intensity $\lambda \sigma(x)$.

Finally, we conclude our discussion about Poisson processes with yet another important property i.e., the Poisson property remains invariant w.r.t random independent translations . It is easy to see the time translation of the points by a fixed amount , say s, i.e., $T_n' = T_n + s$ keeps the Poisson property invariant.

Proposition 1.16 *Let $\{T_n\}$ be the points of a Poisson process with intensity λ. Let $\{\sigma_n\}$ be a collection of i.i.d. random variables, with $\mathbb{E}[\sigma_i] < \infty$, and define:*

$$T_n' = T_n + \sigma_n \tag{1.8}$$

Let N_t' be a point process whose points are $\{T_n'\}$. Then N_t' is a Poisson process with rate λ.

Proof. It is important to note that the sequence $\{T_n'\}$ does not have same monotone property as $\{T_n\}$. To simplify the arguments, the let as assume that the distribution of σ_n is discrete , i.e, $\sigma_n \in \{a_1, a_2, \ldots, a_m\}$ with $\mathbb{P}(\sigma_0 = a_i) = p_i$.

Consider the points T_n that are translated by a_i and denote that subsequence of points by $T_{n,i}'$.

Then, the subsequence $T_{n,i}'$ will be a monotone sequence (chosen from $\{T_n\}$ when the corresponding translation is a_i). From the Poisson thinning property, $N_{t,i}'$, the point process defined by $T_{n,i}'$, is just a Poisson process with rate $\lambda_i = \lambda p_i$. From the definition of T_n' we have:

$$T_n' = T_n + \sum_{i=1}^m a_i \mathbb{1}_{[\sigma_n = a_i]}$$

and thus:

$$N_t' = \sum_{i=1}^m N_{t,i}'$$

But by construction the $N'_{t,i}$ are all independent and Poisson and therefore N'_t is Poisson with rate $\sum_{i=1}^{m} \lambda_i = \lambda$.

The case when σ_0 is continuous can be treated by discretizing the range into m values, invoking the above argument for every m and then passing to the limit via weak convergence.   □

Remark 1.17   The proof above is for the 1-dimensional case, however the property holds for multi-dimensional Poisson processes too. The assumption that the random translations are *i.i.d.* is crucial. This is referred to as the invariance of the Poisson distribution to random shifts.

1.3   STOCHASTIC INTENSITIES AND MARTINGALES

We have referred[1] to λ as the intensity of a Poisson process. The notion of an intensity associated with point processes is much more general and referred to as the *stochastic intensity*.

Let us first introduce some concepts needed understand the notion of a stochastic intensity. Let $(\Omega, \mathcal{F}, \mathbb{P})$ denote a probability space. Suppose for each $t \in T$ where T is an index set in $\Re$, we define an increasing family of sub-sigma fields denoted by $\mathcal{F}_t \in \mathcal{F}$ such that $\mathcal{F}_s \subset \mathcal{F}_t$ whenever $s < t$ then $\mathcal{F}_t$ is called a filtration. For example, the σ-field $\mathcal{F}_t^X = \sigma\{X_s, s \leq t\}$ associated with all events generated by a stochastic process $\{X_t\}$ is called a natural filtration and X_t is said to be adapted to $\mathcal{F}_t$, in other words events such as $\{\omega : X_t(\omega) \leq a\} \in \mathcal{F}_t^X$. Of course we can take $\mathcal{F}_t$ to be just an arbitrary filtration to which $X.$ is adapted.

In the context of point processes we will assume all the filtrations concerned are right continuous, i.e., $\mathcal{F}_{t+} = \bigcap_{n \geq 0} \mathcal{F}_{t+\frac{1}{n}} = \mathcal{F}_t$. A process $\{X_t\}_{t \in T}$ is said to be $\mathcal{F}_t$ predictable if X_t is $\mathcal{F}_{t-}$ measurable. So in particular X_{t-} is $\mathcal{F}_t$ predictable. Note X_{t-} is left-continuous by definition. Indeed, all left continuous $\mathcal{F}_t$ adapted processes are $\mathcal{F}_t$ predictable. Predictable processes play a very important role in the stochastic calculus associated with point processes. We now briefly introduce the main results associated with stochastic intensities. Recall a stochastic process X_t is said to be a $\mathcal{F}_t$- sub, super or simply a martingale if for $t \geq s$:

$$\mathbb{E}[X_t|\mathcal{F}_s] \quad \begin{array}{l} \geq \\ = \\ \leq \end{array} \quad \begin{array}{l} X_s \text{ submartingale} \\ X_s \text{ martingale} \\ X_s \text{ supermartingale} \end{array}$$

Note if X_t is a submartingale then $-X_t$ is said to be a supermartingale. It is easy to see that every simple point process $\{N_t\}_{t \geq 0}$ is a sub-martingale w.r.t. its natural filtration.

There is an important result called the Doob-Meyer decomposition that is associated with sub-martingales that forms the basis for defining martingales associated with point processes. The result is given below.

Theorem 1.18   Doob-Meyer Decomposition

[1]This section can be omitted without affecting the readability and comprehension of the remainder of the text.

Let $\{X_t\}$ be a right-continuous non-negative $\mathcal{F}_t$-submartingale. Then there exists a right-continuous $\mathcal{F}_t$ local martingale[2] $\{M_t\}$ and a $\mathcal{F}_t$ predictable increasing process $\{A_t\}$ such that:

$$X_t = M_t + A_t \tag{1.9}$$

Moreover the decomposition is unique.

We can then use (1.9) to construct martingales associated with point processes from which the notion of a stochastic intensity arises.

Definition 1.19   Let $(\Omega, \mathcal{F}, \mathbb{P})$ be a probability space on which a filtration $\mathcal{F}_t$ is defined and let $\{N_t\}$ be a simple point process adapted to $\mathcal{F}_t$. Then, there exists a non-negative predictable increasing process $\{A_t\}$ called the compensator such that $N_t - A_t$ is a $\mathcal{F}_t$-martingale[3].

If A_t is absolutely continuous with respect to Lebesgue measure then $\exists$ a predictable integrable process $\lambda_t \geq 0$ called the *stochastic intensity* such that $A_t = \int_0^t \lambda_s ds$.

Thus, the Poisson process corresponds to a point process whose stochastic intensity λ is non-random and a constant. Indeed, it follows that if a process has stationary independent increments then its intensity must be constant and therefore must be a Poisson process (of course with respect to the given filtration). A non-homogeneous Poisson process is a point process with compensator $A_t = \int_0^t \lambda_s ds$ and has independent but not stationary increments.

The importance of predictability is the following. Let M_t be a right continuous $\mathcal{F}_t$ martingale and $\{X_t\}$ be a $\mathcal{F}_t$ predictable process with $\mathbb{E}[\int_0^T |X_t|^2 dt] < \infty$. Then $\int_0^t X_s dM_s$ is a $\mathcal{F}_t$ martingale. In other words, if we want to integrate an integrand against a martingale, then the resulting quantity will be a martingale if the integrand is predictable.

To understand this issue of predictability better let us consider the following two integrals associated with a Poisson process with intensity λ (viewed as Stiltjes integrals) $\int_0^t N_s dN_s$ and $\int_0^t N_{s-} dN_s$.

The first integral is:

$$\int_0^t N_s dN_s = \sum_{s \leq t} N_s (N_s - N_{s-}) = 1 + 2 + 3 + \cdots + N_t = \frac{N_t(N_t + 1)}{2}$$

and thus:

$$\mathbb{E}[\int_0^t N_s dN_s] = \frac{\lambda^2 t^2}{2} + \lambda t$$

The second integral is

$$\int_0^t N_{s-} dN_s = \sum_{s \leq t} N_{s-}(N_s - N_{s-}) = \sum_{s \leq t} N_s(N_s - N_{s-}) - \sum_{s \leq t}(N_s - N_{s-})^2 = \frac{N_t(N_t - 1)}{2}$$

---

[2] A local martingale is a process $\{M_t\}$ such that for any sequence of finite $\mathcal{F}_t$ stopping times τ_n converging to infinity, the process $M_{\tau_n \wedge t}$ is an integrable martingale.
[3] Strictly a locally square-integrable martingale.

and thus:

$$\mathbb{E}[\int_0^t N_{s-}dN_s] = \frac{\lambda^2 t^2}{2} .$$

On the other hand, both:

$$\mathbb{E}[\int_0^t N_s \lambda ds] = \mathbb{E}[\int_0^t N_{s-}\lambda ds] = \int_0^t \lambda^2 s ds = \frac{\lambda^2 t^2}{2}$$

and thus we see that taking $M_t = N_t - \lambda t$: $\mathbb{E}[\int_0^t N_s dM_s] \neq 0$ while $\mathbb{E}[\int_0^t N_{s-}dM_s] = 0$ and thus only in the second case we have:

$$\mathbb{E}[\int_0^t N_{u-}dM_u|\mathcal{F}_s] = \int_0^s N_{u-}dM_u$$

showing that $\int_0^t N_{u-}dM_u$ is a $\mathcal{F}_t$ martingale.

Clearly, the results are different. In the first case the integrand is only adapted while in the second case it is predictable.

We can now reinterpret the previous results that we have seen before in the context of Campbell's formula, i.e., (1.6). It is just a re-statement of the martingale property, i.e., for any square-integrable $\mathcal{F}_t$ adapted process:

$$\mathbb{E}[\int_0^t f(X_{s-})dN_s] = \mathbb{E}[\int_0^t f(X_s)\lambda_s ds] = \int_0^t \lambda_s \mathbb{E}[f(X_s)]ds$$

Here the martingale is $N_t - A_t = N_t - \int_0^t \lambda_s ds$.

The compensator associated with the sub-martingale is also related to the notion of a quadratic variation process associated with a martingale. We briefly discuss the issue because it plays an important role in extending the SLLN to martingales. Let $\{M_t\}$ be a square-integrable martingale. Then it readily follows that M_t^2 is a non-negative submartingale.

Formally, the quadratic variation of M_t, denoted by $< M >_t$ is a predictable increasing process of bounded variation such that $M_t^2 - < M >_t$ is a $\mathcal{F}_t$-martingale. From the stationary independent increment property of the Poisson process, it can be readily shown that the compensator A_t is actually the quadratic variation of the martingale $N_t - \lambda t$. In general, the quadratic variation of a martingale associated with a point process can be expressed in terms of the compensator as $< M >_t = \int_0^t (1 - \Delta A_s)dA_s$. When A_t is continuous, $\Delta A_t = 0$. Indeed, the crux of Watanabe's theorem is that if the quadratic variation of the martingale associated with the point process is of the form λt then it must be a Poisson process with intensity λ. The role of the quadratic variation process in our context is its use in extending the SLLN to martingales which we state below without proof.

Theorem 1.20 SLLN for square integrable martingales

Let M_t be a square integrable martingale with $< M >_t$ its quadratic variation. Then:

- a) *On the set $\{\omega :< M >_\infty < \infty\}$ the martingale M_t converges to a finite limit, i.e., $\lim_{t\to\infty} M_t < \infty$.*

- b) *On the set $\{\omega :< M >_t \to \infty \text{ as } t \to \infty\}$ we have $\lim_{t\to\infty} \frac{M_t}{<M>_t} = 0$*

One immediate consequence of this result is that $\lim_{t\to\infty} \frac{N_t - \lambda t}{\lambda t} = 0$ $a.s.$ or $\frac{N_t}{t} \to \lambda$ $a.s.$ as was shown earlier via the classical application of the SLLN.

In general, a stochastic intensity need not be defined. It can be shown that if the distribution of the inter-point times possesses a probability density, then a stochastic intensity is always defined but it need not be deterministic. If N_t is a stationary point process whose inter-arrival times possess a density, then its intensity λ_t is stationary process.

The martingale theory of point processes, or more generally *càdlàg* processes (processes that are right continuous with left limits), is very rich but we will not use the martingale properties in any detail in these notes.

1.4   RESIDUAL LIFE AND THE INSPECTION PARADOX

We now see some other important results related to point processes and Poisson processes that have important implications in the analysis of models of networks. In particular, we discuss some important properties of point processes when they are viewed as sequences of points rather than as counting processes.

Consider the following scenario: Let $\{T_n\}$ be the points of a simple point process. Then the counting process, N_t, that we also refer to as the point process is given by:

$$N_t = \sum_{n=1}^{\infty} \mathbb{1}_{[0 < T_n \leq t]} \tag{1.10}$$

Now by definition $T_{N_t} \leq t < T_{N_t+1}$. We think of $t \in [T_{N_t}, T_{N_t+1})$ as an arbitrary *observation point*, i.e., if an observer arrives at t then t must fall in the interval $(T_{N_t}, T_{N_t+1}]$.

Let us first study some properties about the conditional distributions of the points of a Poisson process . We have already seen that the distribution of the inter-point (or inter-arrival) times of a Poisson process is an exponential distribution and because of the stationary independent increment property of the Poisson process they are independent and identically distributed. Now suppose we know that there are n points in $(0, t]$. What can we say about the distribution of the individual points? In the Poisson case we can completely specify the distributions.

Proposition 1.21   *Let $\{T_i\}$ denote the points of a Poisson process with intensity or rate λ. Then conditionally on knowing that $N_t = n$, the points $\{T_k\}_{k=1}^{n}$ are distributed as the order statistics of n-independent random variables that are uniformly distributed in $[0, t]$.*

Proof. Let us find the distribution of T_1 and then T_n where the association with the minimum and maximum of identically distributed uniform random variables is direct.

$$
\begin{aligned}
\mathbb{P}(T_1 > s | N_t = n) &= \frac{\mathbb{P}(T_1 > s, N_t = n)}{\mathbb{P}(N_t = n)} \\
&= \frac{\mathbb{P}(T_1 > s, N_t - N_s = n)}{\mathbb{P}(N_t = n)} \\
&= \frac{e^{-\lambda t} \frac{(\lambda(t-s))^n}{n!} e^{-\lambda(t-s)}}{\frac{(\lambda t)^n}{n!} e^{-\lambda t}} \\
&= \left(\frac{t-s}{t}\right)^n \\
&= (F(U > s))^n = \mathbb{P}(\min\{U_1, U_2, \ldots, U_n\} > s)
\end{aligned}
$$

where $\{U_i\}$ are i.i.d. Uniform(0,t) random variables.

Similarly:

$$
\begin{aligned}
\mathbb{P}(T_n \leq s | N_t = n) &= \frac{\mathbb{P}(T_n \leq s, N_t = n)}{\mathbb{P}(N_t = n)} \\
&= \frac{\mathbb{P}(T_n \leq s, N_{t-s} = 0)}{\mathbb{P}(N_t = n)} \\
&= \left(\frac{s}{t}\right)^n = (F(U \leq s))^n = \mathbb{P}(\max\{U_1, \ldots, U_n\} \leq s)
\end{aligned}
$$

The general case for T_k is more complicated but repeating calculations of the type above and using the independent increment property we can show T_k is distributed as the k,-th ordered statistic corresponding to the collection of n i.i..d random Uniform(0,t) random variables $\{U_i\}_{i=1}^n$. □

Remark 1.22   Applying the above result to the case $N(s, t] = 1$, we obtain that conditional on knowing that a point has occurred in $(s, t]$ it is uniformly distributed in $(s, t]$.

We now study the notion of forward (and backward recurrence) times that correspond to how long an arbitrary observer that has arrived between two points of a point process has to wait before the occurrence of the next point (or the much time elapsed since the occurrence of the previous point).

Define the following:

$R(t) = T_{N_t+1} - t$: the time the observer has to wait until next point after its arrival (at t). $R(t)$ is called the *forward recurrence time* or *residual or remaining life (or time)*.

$A(t) = t - T_{N_t}$: referred to as the age or the last point before the observer's arrival. $A(t)$ is also called the *backward recurrence time* .

Let $S(t) = T_{N_t+1} - T_{N_t} = A(t) + R(t)$ be the inter-arrival or inter-point time (seen by the arrival at t).

Now suppose N_t is Poisson with rate λ, then we know than $T_{n+1} - T_n$ is exponential with mean $\frac{1}{\lambda}$. Thus, one would expect that: $\mathbb{E}[S(t)] = \frac{1}{\lambda}$. However, this is not so! The fact that the observer arrives at t, biases its observation of the length of the interval—this is called the *inspection paradox*. Let us see why. (Note T_{N_t} is random):

$$
\begin{aligned}
\mathbb{P}(R(t) > x) &= \mathbb{P}(T_{N_t+1} - t > x | T_{N_t+1} > t) \\
&= \mathbb{P}(T_{N_t+1} > t + x | T_{N_t+1} > t) \\
&= \mathbb{P}(T_{N_t+1} - T_{N_t} > t + x - T_{N_t} | T_{N_t+1} - T_{N_t} > t - T_{N_t}) \\
&= \int_0^t \frac{e^{-\lambda(t+x-s)}}{e^{-\lambda(t-s)}} d P_{T_{N_t}}(s) \\
&= e^{-\lambda x}
\end{aligned}
$$

where we used the fact that $T_{N_t+1} - T_{N_t} | T_{N_t} = s \sim exp(\lambda)$ and $\int_0^t d P_{T_{N_t}}(s) = 1$ since $T_{N_t} \leq t$ by definition.

Thus, what we have shown is that the residual life has the same distribution as the inter-arrival time $T_{n+1} - T_n$. Now since $S(t) = A(t) + R(t)$ and by definition $A(t) \geq 0$ it means that:

$$
\mathbb{E}[S(t)] = \mathbb{E}[A(t)] + \mathbb{E}[R(t)] \geq \frac{1}{\lambda}
$$

so that the length of the interval in which the observer arrives is longer than the typical interval.

Let us see what the mean residual life is when the point process is not Poisson but a general renewal point process. Let $S_n = T_n - T_{n-1}$ be the inter-arrival times of a general renewal point process and suppose S_n are i.i.d. with distribution $F(t)$ with $\int_0^\infty x^2 dF(x) < \infty$. Then, assuming $T_0 = 0$ we can write $T_n = \sum_{k=1}^n S_k$. We assume $\mathbb{E}[S_n] = \frac{1}{\lambda_N}$..

Now define $Z_i(t) = (T_{i+1} - t) \mathbb{1}_{[T_i, T_{i+1})}(t)$. Then:

$$
\int_{T_i}^{T_{i+1}} Z_i(s) ds = \frac{1}{2}(T_{i+1} - T_i)^2 = \frac{S_i^2}{2}
$$

Define $Z(t) = \sum_{i=1}^{N_t} Z_i(t)$. Then by definition $Z(t) = R(t)$:

$$
\frac{1}{t} \int_0^t Z(s) ds = \frac{1}{t} \sum_{i=1}^{N_t} \frac{S_i^2}{2} + \frac{1}{t} \int_{T_{N_t}}^t (T_{N_t+1} - s) ds
$$

Now second term on the r.h.s. is smaller than $\int_{T_{N_t}}^{T_{N_t+1}} Z_i(s) ds = \frac{S_{N_t}^2}{2}$

Now letting $t \to \infty$ we have:

$$
\frac{N_t}{t} \cdot \frac{\sum_{k=1}^{N_t} S_k^2}{2(N_t)} + \frac{S_{N_t}^2}{2t}
$$

goes to $\lambda_N \frac{1}{2} \mathbb{E}[S^2]$ since $\frac{S_{N_t}^2}{t}$ goes to zero since S is assumed to have finite second moment.

However,

$$\lim_{t\to\infty} \frac{1}{t} \int_0^t Z(s)ds = \mathbb{E}[R]$$

and noting that $\lambda_N = \frac{1}{\mathbb{E}[S]}$ we have the main result for the mean residual time as:

$$\boxed{\mathbb{E}[R] = \frac{1}{2}\frac{\mathbb{E}[S^2]}{\mathbb{E}[S]}} \qquad (1.11)$$

Some examples

Let us compare the mean residual times for different inter arrival time distributions with the same mean $\frac{1}{\lambda}$.

1. S is deterministic of length $\frac{1}{\lambda}$.

   Then, $\mathbb{E}[S^2] = \frac{1}{\lambda^2}$ and

   $$\boxed{\mathbb{E}[R] = \frac{1}{2\lambda} = \frac{1}{2}\mathbb{E}[S]}$$

2. S is uniform in $[0, \frac{2}{\lambda}]$.

   We see that $\mathbb{E}[S^2] = \frac{4}{3\lambda^2}$ and

   $$\boxed{\mathbb{E}[R] = \frac{2}{3\lambda} = \frac{2}{3}\mathbb{E}[S]}$$

3. S is exponential with mean $\frac{1}{\lambda}$.

   This is the Poisson case and gives, $\mathbb{E}[S^2] = \frac{2}{\lambda^2}$ and

   $$\boxed{\mathbb{E}[R] = \frac{1}{\lambda} = \mathbb{E}[S]}$$

It can be seen that the higher the variability (as measured by the second moment) the higher the residual time, the deterministic case being the smallest.

1.5   EVENT AND TIME AVERAGES (EATA)

We conclude our study with a discussion of the event and time averages problem that leads to the extremely important result of PASTA (Poisson Arrivals See Time Averages) and the important notion of Palm probabilities.

Let $\{X_t\}$ be a stationary ergodic process with $\mathbb{E}[X_t] < \infty$. Then the ergodic theorem states that:

$$\lim_{t \to \infty} \frac{1}{t} \int_0^t f(X_s)ds = \mathbb{E}[f(X_0)]$$

(1.12)

for all bounded functions $f(.)$. Now suppose we observe X_t at times $\{T_i\}$ that correspond to a stationary and ergodic point process N_t and we compute the average over the number of samples we obtain. Define the following if it exists:

$$\mathbb{E}_N[f(X_0)] \stackrel{d}{=} \lim_{N \to \infty} \frac{1}{N} \sum_{k=1}^N f(X_{T_k}) = \lim_{t \to \infty} \frac{1}{N_t} \int_0^t f(X_s)dN_s$$

(1.13)

where we note that $\int_0^t f(X_s)dN_s = \sum_{k=1}^{N_t} f(X_{T_k})$ by the definition of the Stieltjes integral . This is called an event average as we have computed the limit over sums taken at discrete points corresponding to the arrivals of the point process. This arises in applications when we can observe the contents of a queue at arrival times or at times when customers depart but not at all times.

A natural question is: Is $\mathbb{E}_N[f(X_0)] = \mathbb{E}[f(X_0)]$? The answer is no, in general. This is because the limit of the event averages is computed under a probability distribution that corresponds to event or arrival times (that are special instants of time) that are special. This is called a *Palm* probability or distribution. This differs from the time-stationary distribution (of the process $\{X_t\}$). Since this is an important concept (often referred to in queueing literature as Arrival Distributions) we discuss this idea in a purely ergodic framework that sheds some light from an operational point of view.

To extend the framework from Poisson and renewal processes to general stationary point processes we need to impose some mathematical structure, in particular an ergodic framework. Let $(\Omega, \mathcal{F}, P)$ be a probability space with a so-called flow (or shift) operator θ_t be defined with the property that $X_t(\omega) = X_0(\theta_t\omega)$. Moreover, let θ_t be measure preserving i.e., $\mathbb{P} \circ \theta_t(.) = \mathbb{P}(.)$. In that case, $X_t(\omega)$ is said to be a stationary process *consistent* with respect to θ_t. What this means is that the origin of time 0 can be arbitrary since the process has the same probabilistic behavior.

Let $\ldots < T_{-1} < T_0 \leq 0 < T_1 < T_2, \ldots$[4] be a stationary sequence of points, i.e., $T_i - T_{i-1}$ are identically distributed but not necessarily independent. Let $N_t = \sum_n \mathbf{1}_{[0 \leq T_n \leq t]}$ be the counting or point process associated with the $\{T_i\}'s$. This counting process is commonly referred to as a point process. The point process N_t is said to be a stationary point process.

Let $(\Omega, \mathcal{F}, \mathbb{P})$ be a probability space with $\{\theta_t\}$ the shift operator defined as above. Let N_t be a stationary ergodic point process (consistent w.r.t. θ_t). Then for any $A \in \mathcal{F}$ we can define (when

---

[4]Stationary processes and sequences are defined on a doubly infinite interval even if N_t is defined on $[0, \infty)$ and we will work on the positive interval.

the limits exist):

$$\mathbb{P}(A) = \lim_{t \to \infty} \frac{1}{t} \int_0^t \mathbb{1}(\theta_s \circ \omega \in A) ds$$

$$\mathbb{P}_N(A) = \lim_{t \to \infty} \frac{1}{N_t} \int_0^t \mathbb{1}(\theta_s \circ \omega \in A) dN_s = \lim_{t \to \infty} \frac{1}{N_t} \sum_{k=1}^{N_t} \mathbb{1}(\theta_{T_k} \circ \omega \in A)$$

where we have used definition of the Stieltjes integral in the last step. The natural question is what is the relationship between $\mathbb{P}(.)$ and $\mathbb{P}_N(.)$? To better understand this issue let us first study it in the discrete-time setting. The interpretation of $\theta_s \circ \omega$ is that of a sample point at time s.

Let $\{\xi_n\}$ be a stationary sequence (identically distributed) of random variables taking values in $\{0, 1\}$. Let T_n be a set of integer valued r.v. defined as those instants n when $\xi_n = 1$ and let $\lambda_N = \mathbb{P}(\xi_n = 1)$ and we adopt the convention $\ldots < T_{-1} < T_0 \le 0 < T_1 < \ldots$. Let $N[0, n] = \sum_{m=0}^n \mathbb{1}_{[\xi_m=1]}$ denote the point process that counts how many points have occurred in $\{0, 1, 2, \ldots, n\}$. For simplicity we denote $N[0, n]$ as N_n.

Let $\{X_n\}$ be a stationary stochastic sequence and for any $A \in \mathcal{F}$ define :

$$\begin{aligned}
\mathbb{P}_N(X_0 \in A) &= \mathbb{P}(X_n \in A | \xi_n = 1) \\
&= \frac{\mathbb{P}(X_n \in A, \xi_n = 1)}{\mathbb{P}(\xi_n = 1)} \\
&= \frac{\mathbb{P}(X_0 \in A, N[0, 0] = N(\{0\}) = 1)}{\lambda_N} \qquad (1.14)
\end{aligned}$$

Then $P_N(.)$ is called the Palm probability and just corresponds to the probability of an event at time 0 given that a point occurs at 0.

Another way of stating the above result is the following when the underlying processes are ergodic:

$$\begin{aligned}
\lambda_N \mathbb{P}_N(X_0 \in A) &= \mathbb{E}[\mathbb{1}_{[X_0 \in A, \xi_0=1]}] \\
&= \lim_{n \to \infty} \frac{1}{n} \sum_{k=1}^{N_n} \mathbb{1}_{[X_{T_k} \in A]} \\
&= \lim_{n \to \infty} \frac{N_n}{n} \frac{1}{N_n} \sum_{k=1}^{N_n} \mathbb{1}_{[X_{T_k} \in a]}
\end{aligned}$$

From the definition of Palm probabilities, we see immediately that $\mathbb{P}_N(T_0 = 0) = \mathbb{P}_N(\xi_0 = 1) = 1$ or there is a point at the origin w.p.1. Let us now obtain the equivalent of Campbell's formula in this setting. Let $\{X_n, \xi_n\}$ be a jointly stationary sequence with $\xi_n \in \{0, 1\}$ and $X_n \in \mathfrak{R}$, then:

$$\begin{aligned}
\mathbb{E}[\sum_n X_{T_n} \mathbb{1}_{[0 \le T_n \le k]}] &= \sum_{n=0}^k \mathbb{E}[X_n \mathbb{1}_{[\xi_n=1]}] \\
&= (k+1)\lambda_N \mathbb{E}_N[X_0]
\end{aligned}$$

by the definition of $\mathbb{P}_N(.)$ above. The only difference between this formula and Campbell's formula (1.6) is that the underlying point process was a continuous-time Poisson process in that context, and therefore instead of computing the mean with respect to $\mathbb{P}_N$, the expectation there was computed with respect to the probability $\mathbb{P}$, an issue that we will see a bit later in the context of the PASTA property.

We have still not answered the question of the relationship between the two probability measures, $\mathbb{P}(.)$ and $\mathbb{P}_N(.)$. This is given by the inversion formula.

Proposition 1.23 Discrete-time Palm Inversion Formula
  Let $\{X_n, \xi_n\}$ be jointly stationary and ergodic sequences on $(\Omega, \mathcal{F}, \mathrm{P})$ and $X_n \in \mathfrak{R}$ and $\xi_n \in \{0, 1\}$.
Then:

$$\mathbb{E}[X_n] = \mathbb{E}[X_0] = \lambda_N \mathbb{E}_N[\sum_{k=0}^{T_1-1} X_k] \tag{1.15}$$

where T_1 denotes the first index $n > 0$ when $\xi_n = 1$.

Proof. We will prove this result by an ergodic argument since this clearly highlights the interpretation of time and event averages of the stationary probabilities and the Palm probabilities.

$$\mathbb{E}[X_0] = \lim_{n \to \infty} \frac{1}{n+1} \sum_{k=0}^{n} X_k = \lim_{n \to \infty} \left[\frac{N_n}{n+1} \frac{1}{N_n} \sum_{k=0}^{N_n} Y_k + \frac{1}{n+1} \sum_{k=T_{N_n}+1}^{n} X_k \right]$$

where $Y_k = \sum_{j=T_{N_k}}^{T_{N_{k+1}}-1} X_j$.

All we have done is that we have replaced the sum of X_k over k to sums of random variables over *cycles* that are of length $[T_{N_k}, T_{N_{k+1}-1}]$. The last term is just to account for the difference between T_{N_n} and n. Now as $n \to \infty$ the first term on the right-hand side converges to $\lambda_N \mathbb{E}_N[\sum_{k=0}^{T_1-1} X_k]$ and the second term goes to zero since it is less than the sum of X_k over one cycle (i.e. consists only of a finite number of terms). $\qquad\square$

Let us now return to the continuous-time case keeping in mind the notion of Palm probabilities as conditioning on the event that a point occurs. In continuous time this has a probability of 0 and it is because of this *conditioning on an event of measure 0* that we need more sophisticated machinery. However, in the ergodic context, we can always retain the interpretation of that Palm probabilities as being averages taken over events. We begin by stating and proving the Palm inversion formula to show how the discrete-time techniques directly carry over.

Proposition 1.24 Palm Inversion Formula

Let $\{X_t\}$ be a stationary process that is consistent w.r.t. θ_t and $\{N_t\}$ be a stationary simple point process also consistent w.r.t. θ_t and $\mathbb{E}[X_0] < \infty$. Then:

$$\boxed{\mathbb{E}[f(X_0)] = \lambda_N \mathbb{E}_N[\int_0^{T_1} f(X_s)ds]} \tag{1.16}$$

Proof. Keeping in the spirit of our discussion so far, let us assume both N and X are ergodic and hence: $\lim_{t \to \infty} \frac{N_t}{t} = \lambda_N$ $a.s.$ by definition of λ_N as the average intensity.

Without loss of generality let us take $f(X_t) = X_t$. Then from the ergodicity of X we have:

$$\mathbb{E}[X_0] = \lim_{t \to \infty} \frac{1}{t} \int_0^t X_s ds$$
$$= \lim_{t \to \infty} \frac{N_t}{t} \frac{1}{N_t} \sum_{k=1}^{N_t} Y_{T_k} + \lim_{t \to \infty} \frac{1}{t} \int_{T_{N_t}}^t X_s ds$$

where $Y_{T_k} = \int_{T_{k-1}}^{T_k} X_s ds$ and $T_0 = 0$ by convention.

Now as $t \to \infty$ the last term goes to 0 while $\frac{1}{N_t} \sum_{k=1}^{N_t} Y_{T_k}$ goes to $\mathbb{E}_N[\int_0^{T_1} X_s ds]$ by definition of Y_{T_k} and hence the result follows. $\square$

Some remarks are in order before we proceed.

Remark 1.25

- By definition of $\mathbb{P}_N$ we have $\mathbb{P}_N(T_0 = 0) = 1$

- From the inversion formula (by taking $f(.) = 1$) we see

$$\mathbb{E}_N[T_1] = \mathbb{E}_N[T_n - T_{n-1}] = \frac{1}{\lambda_N}$$

- From the definition of Palm probability and the consistency of N_t w.r.t. θ_t we have $\mathbb{P}_N \circ \theta_{T_N}(.) = \mathbb{P}_N(.)$ i.e., $\mathbb{P}_N$ is invariant w.r.t shifts θ_{T_n}.

- The interpretation of the expectation w.r.t. $\mathbb{P}_N$ is that $\mathbb{E}_N[f(X_0)] = \mathbb{E}[f(X_t)|\Delta N_t = 1]$, i.e., we are conditioning at an instant at which a point occurs.

- From the definition of the Palm probability it can be seen that if $\{X_t\}$ is a stationary process and N_t is a stationary point process (consistent w.r.t. θ_t) then:

$$\mathbb{E}[\int_0^t X_s dN_s] = \lambda_N t \mathbb{E}_N[X_0]$$

Let us return to the question as to when the two probability measures coincide. They coincide when the times correspond to arrival times of a Poisson process. This is the important property of PASTA , first shown by Wolff and later generalized by Bremaud et. al. Here is a heuristic proof of why it holds.

Proposition 1.26 PASTA

Let X_t be a stationary ergodic process and let $\mathcal{F}_t$ denote an increasing information flow (i.e., $\mathcal{F}_s \subset \mathcal{F}_t$, $s < t$)[5] such that we can determine X_s, $s \le t$ if we know $\mathcal{F}_t$. Now let N_t be a Poisson process with rate λ also determined by $\mathcal{F}_t$ i.e., the event $\{N_t = n\} \in \mathcal{F}_t$.

Let $\{T_n\}$ denote the points of N_t. Then:

$$\lim_{N \to \infty} \frac{1}{N} \sum_{k=1}^{N} f(X_{T_k-}) = \boxed{\mathbb{E}_N[f(X_{0-})] = \mathbb{E}[f(X_0)]} = \lim_{t \to \infty} \frac{1}{t} \int_0^t f(X_s)ds \qquad (1.17)$$

Proof. (Heuristic). A rigorous proof needs the martingale property (done below). Here is a heuristic proof on why the Poisson property is needed. Without loss of generality, we take $f(x) = x$.

$$\mathbb{E}_N[X_{t-}] = \mathbb{E}[X_{T_n-}] = \mathbb{E}[X_{t-}|\Delta N_t = 1]$$

where $\Delta N_t = N_t - N_{t-}$ denotes a jump or a point occurring. Note $\Delta N_t = 1$ when $t = T_n$, $n = 1, 2, \cdots$, and 0 otherwise.

Now:

$$\begin{aligned}
\mathbb{E}[X_{t-}|\Delta N_t = 1] &= \frac{\mathbb{E}[X_{t-}\mathbb{1}_{[\Delta N_t=1]}]}{P(\Delta N_t = 1)} \\
&= \frac{\mathbb{E}[X_{t-}]\mathbb{E}[\mathbb{1}_{[\Delta N_t=1]}]}{P(\Delta N_t = 1)]} \\
&= \mathbb{E}[X_{t-}] = \mathbb{E}[X_0]
\end{aligned}$$

Here the key is that the second step follows from the independent increment property of N_t because it is Poisson and so $\Delta N_t = \lim_{\delta \to 0} N_{t-+\delta} - N_{t-}$ is independent of X_{t-}. Note because of Watanabe's theorem a Poisson process is the only point process that has the stationary independent increment property. $\qquad \square$

---

[5]The right term is filtration.

Proof. (Rigorous using the martingale SLLN[6]). Let X_t be $\mathcal{F}_t$ adapted and N_t be a $\mathcal{F}_t$ adapted Poisson process with intensity $\lambda > 0$. By definition:

$$\mathbb{E}_N[X_{0-}] = \lim_{t \to \infty} \frac{1}{N_t} \int_0^t X_{s-} dN_s$$

Assume that $\sup_t \mathbb{E}[X_t]^2 < \infty$. Define the Poisson martingale $M_t = N_t - \lambda t$.

Then $Y_t = \int_0^t X_{s-} dM_s$ is a martingale with quadratic variation $\int_0^t X_{s-}^2 \lambda ds \to \infty$ *as* $t \to \infty$.

From ergodicity:

$$\lim_{t \to \infty} \frac{1}{t} \int_0^t X_s^2 ds = \mathbb{E}[X_0^2] \tag{1.18}$$

Therefore, using the martingale SLLN we have:

$$\frac{(\int_0^t X_{s-} dM_s)}{\int_0^t X_s^2 \lambda ds} \to 0 \ as \ t \to \infty$$

which is equivalent to:

$$\lim_{t \to \infty} \frac{N_t \int_0^t X_{s-} dN_s}{N_t \int_0^t X_s^2 \lambda ds} - \frac{t \int_0^t X_s \lambda ds}{t \int_0^t X_s^2 \lambda ds} = 0$$

or by appropriate grouping of terms and (1.18) we have:

$$\mathbb{E}_N[X_{0-}] = \mathbb{E}[X_0]$$

□

Remark 1.27   In the result above the l.h.s. is stated as $\mathbb{E}_N[X_{0-}]$. But if X_t and N_t do not have common jumps or X_t is continuous, then $\mathbb{E}_N[X_{0-}] = \mathbb{E}_N[X_0]$ which is how usually PASTA is stated. Also the martingale SLLN proof can be replaced by a more direct proof using only an integrability (first moment) assumption on X_t and the martingale property.

In light of the definition of the Palm probability let us now re-visit the inspection paradox that we discussed in the special case of renewal processes.

Define $R_t = T_1 - t$ and $A_t = t - T_0$ when $t \in [T_0, T_1)$. Then:

$$\mathbb{E}[A_t + R_t] = \mathbb{E}[T_1 - T_0] = \lambda_N \mathbb{E}_N[\int_0^{T_1} (A_s + R_s) ds]$$

But $\int_0^{T_1} A_s ds = \int_0^{T_1} R_s ds = \frac{T_1^2}{2}$. Therefore:

$$\mathbb{E}[T_1 - T_0] = \lambda_N \mathbb{E}_N[(T_1 - T_0)^2]$$

[6]This proof can be skipped if Section 1.3 has been skipped

Here we have used use the fact that under $\mathbb{P}_N$, $T_0 = 0$.

Now noting that $\mathbb{E}_N[T_1 - T_0] = \frac{1}{\lambda_N}$ and $\mathbb{E}_N[(T_1 - T_0)^2] = (\mathbb{E}_N[T_1 - T_0])^2 + var_N(T_1 - T_0)$ (where var_N denotes the variance taken under $\mathbb{P}_N$) we obtain that:

$$\mathbb{E}[T_1 - T_0] \geq \frac{1}{\lambda_N}$$

since $var_N(T_1 - T_0) \geq 0$ by definition of the variance. The exact amount by which the time-stationary mean of the interval is over estimated is $\lambda_N var_N(T_1 - T_0)$.

When N_t is Poisson, by PASTA we have $\mathbb{P}_N = \mathbb{P}$ and we see that in the stationary case:

$$\mathbb{E}[T_{N_t+1} - T_{N_t}] = \frac{2}{\lambda_N}$$

implying that the interval is twice as long on average!

We conclude our discussion with two important formulae that are very useful in calculations, especially in the context of queueing systems where in addition to a so-called driving point process associated with the input there are secondary point processes that can be associated with certain events of interest and it might be easier to compute with respect to the Palm distribution associated with the derived point process. The first is called Neveu's exchange formula . A proof will be given in the next chapter.

Proposition 1.28   Neveu's Exchange Formula

Let N and N' be two stationary point processes and $\{X_t\}$ be a stationary process all defined on $(\Omega, \mathcal{F}, \mathbb{P})$ that are consistent w.r.t. θ_t. Let λ_N and $\lambda_{N'}$ be the average intensities of N and N', respectively, and let $\{T_n\}$ and $\{T'_n\}$ denote their points and $\lim_{n\to\infty} T_n = T'_n = \infty$ a.s.. Then for any measurable function such that $\mathbb{E}_N[f(X_0)] < \infty$:

$$\boxed{\lambda_N \mathbb{E}_N[f(X_0)] = \lambda_{N'} \mathbb{E}_{N'}[\int_0^{T'_1} f(X_s)dN_s]} \tag{1.19}$$

where $\mathbb{E}_N[.]$ and $\mathbb{E}_{N'}[.]$ denote the expectations calculated w.r.t. the respective Palm probabilities.

The second formula of interest is one that allows us to relate computations between the Palm measure and the stationary measure when we know the stochastic intensity of the underlying point process. As such it is a generalization of the EATA idea when the underlying point process is not Poisson. In fact, the importance of this result is that it ties in the Palm theory to the martingale theory of point processes. This result is called Papangelou's formula which is stated and proved exploiting the martingale SLLN.

Proposition 1.29   Papangelou's Formula

Let $(\Omega, \mathcal{F}, \mathbb{P})$ be a probability space that carries a filtration $\mathcal{F}_t$. Let $\{X_t\}$ be a $\mathcal{F}_t$-adapted stationary and ergodic process that is jointly stationary with an ergodic point process $\{N_t\}$ that possesses a $\mathcal{F}_t$-stochastic

intensity given by $\lambda_t = \lambda(\theta_t \omega)$. Let $\mathbb{P}_N$ and $\mathbb{P}$ denote the Palm and stationary distributions for X and $\lambda_N = \mathbb{E}[\lambda_0]$ denote the average intensity.
    Then:

$$\boxed{\lambda_N \mathbb{E}_N[X_{0-}] = \mathbb{E}[X_0 \lambda_0]} \tag{1.20}$$

Proof. We will use ergodicity and the martingale SLLN to give a proof. For this let us assume as in the proof of PASTA that $\sup_t \mathbb{E}[|X_t|^2] < \infty$. Since by assumption λ_t is the intensity and N_t is ergodic it follows that $\lim_{t\to\infty} \frac{1}{t} \int_0^t \lambda_s ds = \lambda_N$ a.s..
    As in the proof of PASTA, using the SLLN as well as ergodicity, we can show that

$$\lim_{t\to\infty} \frac{1}{t} \int_0^t X_{s-}(dN_s - \lambda_s ds) = 0$$

and hence once again re-writing the above we have:

$$\lim_{t\to\infty} \left\{ \frac{N_t}{t} \frac{\int_0^t X_{s-} dN_s}{N_t} - \frac{1}{t} \int_0^t X_s \lambda_s ds \right\} = 0$$

Noting by ergodicity $\lim_{t\to\infty} \frac{N_t}{t} = \lambda_N$ and the definition of Palm probabilities, the result follows.
□

Remark 1.30   Papangelou's formula can be re-written as:

$$\boxed{\mathbb{E}_N^0[f(X_{0-})] = \mathbb{E}[f(X_0)] + \frac{cov(f(X_0), \lambda_0)}{\mathbb{E}[\lambda_0]}} \tag{1.21}$$

where $X_t = X_0(\theta_t \omega)$ is a stationary process and $\lambda_t = \lambda_0(\theta_t \omega)$ is the stochastic intensity of N.

To see how this result follows: Indeed, for any $X_0 \in \mathcal{F}_{0-}$, Papangelou's formula reads as:

$$\lambda_N \mathbb{E}_N[f(X_0)] = \mathbb{E}[f(X_0)\lambda_0]$$

    Now:

$$cov(f(X_0), \lambda_0) = \mathbb{E}[f(X_0)\lambda_0] - \mathbb{E}[f(X_0)]\mathbb{E}[\lambda_0]$$

Noting that $\lambda_N = \mathbb{E}[\lambda_0]$ we have:

$$\mathbb{E}_N[f(X_0)] = \frac{cov(f(X_0), \lambda_0)}{\mathbb{E}[\lambda_0]} + \mathbb{E}[f(X_0)]$$

From here we immediately obtain the following result: if N is Poisson with (constant) intensity λ we see $cov(f(X_0), \lambda_0) = 0$ and thus:

$$\mathbb{E}_N[f(X_{0-})] = \mathbb{E}[f(X_0)]$$

establishing PASTA. The equation (1.21) also shows that PASTA thus holds if λ_t and $f(X_t)$ are uncorrelated under $\mathbb{P}$, i.e., it can be extended to the doubly stochastic poisson case provided the intensity and the process $\{X_t\}$ are uncorrelated as for example in the situation when a continuous-time stationary stochastic process is sampled at times of a doubly stochastic Poisson process independent of it.

This concludes our brief look at point processes and the definition of probability measures associated with them. We will now study another type of process that plays an important role in traffic modeling and how one might define the idea of arrival distributions that can be loosely termed as fluid Palm distributions.

1.6   FLUID TRAFFIC ARRIVAL MODELS

Another class of traffic models that are of importance in the context of today's networks are so-called fluid traffic models. Here the arrivals are continuous in time and are defined by an instantaneous rate rather than an average rate as in the case of point process arrivals. This is because at the speeds of today's networks the granularity of discrete arrivals of packets can be ignored and instead the arrival pattern seen is an arrival of a continuous stream of bits as depicted in Figure 1.1. Once again, in order to develop useful results we need to impose some hypothesis on the arrivals.

Fluid queueing models are of many types. For example, an important class of fluid models that are used for obtaining worst case estimates of network performance are so-called regulated flow arrival models or traffic envelopes. These are deterministic bounds on the cumulative volume of arrivals in a given time interval. The most common of these types of fluid models are so-called *leaky bucket* models. These are specified by three parameters (π, ρ, σ) with $\rho < \pi$ where π represent the peak rate of arrival of bits, ρ is a bound on the long-term average arrival rate, and σ is the maximum burst size of the arrival process. In particular, the cumulative input in any interval of length t satisfies,

$$A(0, t) \leq \min\{\pi t, \rho t + \sigma\}, \quad \forall t > 0 \tag{1.22}$$

From the definition we see that: $\sup_t \frac{A(0,t)}{t} = \pi$, $\lim_{t \to \infty} \frac{A(0,t)}{t} \leq \rho$ and σ represents the maximum instantaneous burst size of arrivals.

The regulated traffic model is specified in terms of a deterministic bound on the total number of arrivals and is useful when we need insights on worst-case behavior. The use of regulated traffic models for studying the performance of networks is termed as *network calculus* . One of the principal goals of this book is to develop tools that allow us to understand the statistical behavior of communication networks based on statistical models for the traffic. Readers interested in the network calculus approach should refer to the books by Chang, and LeBoudec *et al.* listed at the end of the chapter.

Definition 1.31   A stochastic process $A_t : \Re \to \Re^+$ is said to be a stationary fluid arrival process if A_t is continuous and increasing in t and the increments $Y_s(t) = A_{t+s} - A_s$ is a stationary process in s for fixed t, in other words the distribution of $A_{t+s} - A_s$ only depends on t.

A particular subclass of fluid arrival models, the so-called ON-OFF arrival model plays an important role in modeling traffic arrivals in high-speed networks. It is specified by sequences of r.v's $\{O_i\}$ and $\{S_i\}$ which are called periods on activity (ON periods) and periods of silence or inactivity (OFF periods) and characterized by:

$$A_{Z_i+t} - A_{Z_i} = F_i(t - Z_i)\mathbb{1}_{[t\in[Z_i, Z_i+\sigma_i)]} \tag{1.23}$$

where $Z_i = \sum_{j=0}^{i-1}(\sigma_j + S_j)$ with the convention $\sigma_0 = S_0 = 0$. In other words, during an activity period σ_i the source increases as $F_i(t - .)$ and during a silent period the input is identically 0. The particular case of interest is when the rate (derivative of A_t) is constant during an active period and given by r_i. The interpretation of the ON period is it represents the *burst* length or duration in which arrivals take place.

Of course to obtain a statistical description we need to define how $F_i(.)$, σ_i and S_i behave. One particularly simple model is of exponentially distributed ON and OFF periods with a constant rate of bits when in the ON state. A typical realization of such a process is shown in Figure 1.3.

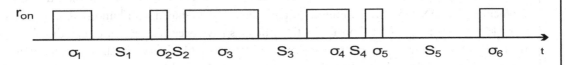

Figure 1.3: Realization of an ON-OFF fluid source.

Definition 1.32   A fluid arrival process is said to be a constant rate exponential arrival process if the rate r_{ON} during the ON periods σ_i is constant and σ_i and S_i are i.i.d. exponentially distributed with parameters λ_{ON} and λ_{OFF} respectively.

From the definition of an ON-OFF source the following is immediate:

Lemma 1.33   *Let A_t be a stationary fluid ON-OFF process with constant rate when in the ON state, then the average rate of the source is given by:*

$$\boxed{r_{avg} = r_{ON}\frac{\mathbb{E}[\sigma_i]}{\mathbb{E}[\sigma_i] + \mathbb{E}[S_i]}} \tag{1.24}$$

Remark 1.34   For exponential ON-OFF processes:

$$\boxed{r_{avg} = r_{ON} \frac{\lambda_{OFF}}{\lambda_{ON} + \lambda_{OFF}}} \qquad (1.25)$$

By changing the behavior of the ON and OFF distributions of a fluid ON-OFF source we can model a wide variety of source behaviors. One of the important behaviors is the so-called property of long-range dependence. Long-range dependence (LRD) relates to phenomena where correlations between a process at different times persists for a long time that is mathematically represented by the non-integrability of the covariance function. A related concept is that of *self-similarity*. Self-similarity relates to the phenomenon where the statistics of the process are *scale–invariant*, i.e., they look the same at all time scales. In Internet traffic measurements made in the 1990's it was observed that traffic exhibited both LRD and self-similar characteristics. This led to the search of traffic models that could capture such behavior. Strictly speaking LRD is more natural since self-similarity is more associated with fractional Brownian motion (fBm) type models (for tractability) that are more difficult to justify. LRD behavior can be exhibited by considering fluid ON-OFF models where the ON distribution has a power-law tail distribution. More precisely, we can consider models where the ON distribution is such that $\mathbb{P}(\sigma_i > x) \sim C x^{-(1+\alpha)}$ where $\alpha > 0$. If $\alpha < 1$ then the second moment of the ON time will be infinite (often called a heavy-tailed distribution). Although such a model can be thought of as unrealistic, these models have been used because such distributions (called Pareto distributions) occur in many applications where file sizes with Pareto distributions are not unrealistic. Models with $\alpha \in (1, 2)$, i.e., having finite 1st and 2nd moments can also result in LRD behavior and these are more realistic. These notes will not deal with such phenomena and an interested reader can refer to the bibliography given at the end.

Remark 1.35   One can relate the regulated fluid arrival model to the ON-OFF fluid arrival model by considering the following periodic ON-OFF model of period $T > \frac{\sigma}{\pi - \rho}$ with rates given by:

$$
\begin{aligned}
r_t &= \pi, & t &\in (0, \frac{\sigma}{\pi - \rho}) \\
&= \rho, & t &\in (\frac{\sigma}{\pi - \rho}, T - \frac{\sigma}{\rho}) \\
&= 0, & t &\in (T - \frac{\sigma}{\rho}, T)
\end{aligned}
$$

It can then be checked that $A(0, t) = \int_0^t r_s ds \leq \min\{\pi t, \rho t + \sigma\}$. Randomness can be introduced by randomizing the phase of the process.

We conclude this introduction by introducing the notion of a fluid Palm distribution. As before we will restrict ourselves to the ergodic case.

Definition 1.36 Let A_t be a stationary, ergodic continuous increasing process with stationary increments that is consistent with a flow θ_t on $(\Omega, \mathcal{F}, \mathbb{P})$ with $\lim_{t \to \infty} A_t \to \infty$ *a.s.*. Define the following limit:

For any $C \in \mathcal{F}$,

$$\mathbb{P}_A(C) = \lim_{t \to \infty} \frac{1}{A_t} \int_0^t \mathbb{1}(\theta_s \omega \in C) dAs \qquad (1.26)$$

If $\{X_t\}$ is a stationary and ergodic process that is also consistent with θ_t then

$$\mathbb{E}_A[X_0] = \lim_{t \to \infty} \frac{1}{A_t} \int_0^t X_s dA_s \qquad (1.27)$$

$\mathbb{P}_A(.)$ is called the fluid Palm measure associated with A and $\mathbb{E}_A[.]$ denotes the corresponding expectation under $\mathbb{P}_A$

The interpretation of $\mathbb{P}_A$ is that it is the probability associated with an arriving *bit*. There is no simple inversion formula for fluid Palm measures and the relationship between the fluid Palm measure and the time-stationary probability measure is problem dependent as will be shown later.

In the next chapter we will study the application of these results to queues and especially the interpretation of quantities of interest to determining the performance of networks.

NOTES AND PROBING FURTHER

The material covered in this chapter forms the basis of modeling and traffic engineering in networks. The choice of the issues that have been presented is one that is motivated by the applications that will be discussed in the subsequent chapters. Since the treatment has only been suggestive of the tools and results, the reader is recommended to consult the books and articles given below to probe further. The references are indicative and not exhaustive.

The basic background required is one of stochastic processes.

BOOK REFERENCES

G. R. Grimmett and D. R. Stirzaker, *Probability and Random Processes*, Oxford Science Publ, 1998

This is an excellent book for all the basic probability tools used in this book.

R.W. Wolff; *Stochastic Modelling and the Theory of Queues*, Prentice-Hall, N.J., 1989

This book is a very detailed text for understanding the basics of queues and renewal theory.

F. Baccelli and P. Brémaud; *Elements of Queueing Theory: Palm-Martingale Calculus and Stochastic Recurrences*, 2nd Ed., Springer-Verlag, N. Y., 2005

This is an advanced book that presents a comprehensive look at Palm theory with queueing applications. It is difficult for beginners who do not have a strong background in analysis and measure theoretic probability. However, it is a *tour de force* and highly recommended to readers interested to develop a thorough understanding of the modeling and analysis of queueing systems.

K. Sigman; *Stationary Marked Point Processes: An Intuitive Approach*, Chapman and Hall, NY, 1995.

This book is a very good reference for understanding the basics of the interpretation and operational aspects of Palm theory and stationary point processes in general. The approach is based along an ergodic framework, much as has been done in this chapter.

W. Willinger; *Self-similar Network Traffic and Performance Evaluation*, John Wiley and Sons, N.Y., 2000

This book presents the background on the alternative traffic models that were presented at the end of the chapter.

C-S. Chang; *Performance Guarantees in Communication Networks*, Springer-Verlag, London, 2000.

J-Y. Le Boudec and P. Thiran; *Network calculus. A theory of deterministic queuing systems for the internet.* Lecture Notes in Computer Science, 2050. Springer-Verlag, Berlin, 2001.

These last two books present the network calculus approach based on regulated traffic models for performance analysis. Such an approach provides worst-case estimates of network behavior and also allows us to understand the network dynamics through deterministic tools drawn from (min,+) algebra.

JOURNAL ARTICLES

P. Brémaud, R. Kannurpatti and R. R. Mazumdar; Event and time averages: A review and some generalizations, *Adv. in Appl. Prob*, 24,(1992), pp. 377-411

The complete set of results on PASTA type of issues but needs background as in Baccelli and Brémaud above. In particular, the paper discusses EATA issues via the theory of stochastic intensities that provides connections between the martingale theory of point processes and Palm theory through Papangelou's theorem in much more detail with proofs that are more general than have been presented here.

W. Willinger, W.E. Leland, M.S. Taqqu, and D.V. Wilson, On the Self-Similar Nature of Ethernet traffic, *IEEE/ACM Transactions on Networking*, Vol. 2 (1), 1994, pp. 1-15

This is the paper that first reported the need to move away from conventional point process models to account for LRD and self-similarity based on measurements of Internet traffic. This gave impetus to study queues with fluid inputs in a much more concerted way.

S. Resnick, and G. Samorodnitsky; Limits of on/off hierarchical product models for data transmission, *Ann. Appl. Probab.*, 13 (2003), no. 4, 1355–1398.

This paper shows how LRD effects can be obtained from ON-OFF traffic models based on the distributions of ON and OFF periods.

C. Rainer, and R. R. Mazumdar ; A note on the conservation law for continuous reflected processes and its application to queues with fluid inputs, *Queueing Systems*, Vol. 28, Nos. 1-3,(1998), pp.283-291.

T. Konstantopoulos, and G. Last, On the dynamics and performance of stochastic fluid systems, *J. Appl. Probab.*, 37 (2000), no. 3, 652–667.

The last two papers deal with the definition of fluid Palm measures and how they can be used to analyze ON-OFF types of inputs. In the next chapter we will show how we can perform computations in this framework.

CHAPTER 2

Queues and Performance Analysis

In the previous chapter we saw the modeling of the traffic arrival processes either as point processes or as fluid arrival processes. Because of the stochastic fluctuations of the arrivals at routers and switches, they can become overloaded. If there is buffer space available to store the packets, those that cannot be served are stored in the buffer. The key in network design is to ensure that there are sufficient resources available to process the packets that arrive and that temporary overloads do not result in buffer overflows whereby packets can be lost. The main question is to understand the stochastic behavior of such systems and one of the key notions is that of *stability*. Stability means that on the long run the system behaves fine and the packets that arrive can be processed. However, because of the random arrivals the temporary overloads cause queueing and so it is important to understand the queueing process. We must develop means to estimate (on average) how large the queues can become, if the server has enough capacity or speed, etc. One of the insights necessary is to understand what parameters of the traffic arrival process determine the queueing behavior and performance, and how dependent are the results on the modeling assumptions. In this chapter we will try to answer some of these questions and see some key results that allow us to predict the performance of networks.

It is important to stress that the results in this chapter are not exhaustive. The goal is to provide the right level of modeling and analysis to understand and develop the results required in the sequel.In particular, the approach that has been taken is a *modern* sample-path based approach whereby a lot can be inferred about the queueing relations under very general hypothesis. Of course specializing the results by imposing more conditions allows us to obtain more detailed insights. In particular, the approach followed shows how the various concepts and models introduced in Chapter 1 play a role in the determination of queueing formulae and their interpretation.

2.1 PRELIMINARIES: RATE CONSERVATION LAW (RCL) AND GENERALIZATIONS

The sample-path of queueing processes when arrivals are modeled as point processes is referred to as càdlàg[1] processes (right continuous processes with left-hand limits). This is a consequence of the fact that the arrivals and departures occur at discrete random times that are unpredictable (in general)

---

[1] càdlàg refers to the French abbreviation continue à droite, limites à gauche

and cause discontinuities that are captured by sample-paths that belong to this class of processes. It should be noted that continuous processes are càdlàg. It turns out that the structure of càdlàg processes can be exploited to obtain a very general result that relates the mean of the continuous part to the mean of the jumps or discontinuities through the use of the Palm probability that we introduced in Chapter 1.

We now state some basic properties of càdlàg processes that will be used in this chapter. We omit the proofs as they can be found in any book on real analysis.

Definition 2.1   A càdlàg process $\{X_t\}$ is said to be of bounded variation if for any $[a, b]$ with $|b - a| < \infty$

$$v([a, b]) = \sup_{\mathcal{D}} \sum_{i=0}^{N-1} |X_{t_{i+1}} - X_{t_i}| < \infty \tag{2.1}$$

where $\mathcal{D}$ ranges over all subdivisions $a = t_0 < t_1 < \cdots t_N = b$ for all N.

A càdlàg process of bounded variation can have jumps and we will henceforth assume that on a finite interval there can be only a finite number of such jumps. Roughly speaking, being of bounded variation means that the sum of the magnitude of all jumps in any finite interval is finite and moreover if we move along on the sample-path then the distance covered be finite in a finite interval of time. The models we will deal with will have this property. An example of a continuous process with unbounded variation is Brownian Motion or the Wiener process. Such processes have paths that are so irregular that the length is infinite on any finite interval.

Every càdlàg process $\{X_t\}$ having jump discontinuities can be written as:

$$X_t = X_0 + X_t^c + X_t^d$$

where X_t^c is purely continuous (w.r.t. t) and X_t^d is purely discontinuous.
    Let $\Delta X_t = X_t - X_{t-}$ denote the jump of $\{X_t\}$. Then:

$$X_t^d = \sum_{0 < s \le t} \Delta X_s$$

Furthermore, the continuous part is differentiable and has a so-called right derivative that is defined below.

Definition 2.2   Let $\{X_t\}$ be a càdlàg process of bounded variation. Then the right derivative defined as:

$$X_t^+ = \lim_{\varepsilon \to 0} \frac{X_{t+\varepsilon} - X_t}{\varepsilon} \tag{2.2}$$

exists.

Therefore, we can obtain the following representation for any càdlàg process of bounded variation.

Proposition 2.3   *Let $\{X_t\}$ be a càdlàg process of bounded variation. Then for any $t > 0$:*

$$X_t = X_0 + \int_0^t X_s^+ ds + \sum_{0 < s \leq t} \Delta X_s \tag{2.3}$$

where $\Delta X_s = X_s - X_{s-}$ denotes the jumps or discontinuities of the process.

We now recall another important result associated with càdlàg processes that is referred to as the Lebesgue-Stieltjes Integration by parts formula.

Proposition 2.4   *Let X_t and Y_t be two càdlàg processes of bounded variation. Then:*

$$X_t Y_t = X_0 Y_0 + \int_0^t X_s dY_s + \int_0^t Y_{s-} dX_s \tag{2.4}$$

or equivalently

$$X_t Y_t = X_0 Y_0 + \int_0^t X_s Y_s^+ ds + \int_0^t Y_s X_s^+ ds + \sum_{0 < s \leq t} \Delta X_s \Delta Y_s \tag{2.5}$$

Define $N_t = \sum_{0 < s \leq t} \mathbb{1}_{[X_s \neq X_{s-}]}$. Then N_t is the point process that counts the jumps of X_t and by definition is a simple point process. Then by the definition of the Stieltjes integral we can write (2.3) as:

$$X_t = X_0 + \int_0^t X_s^+ ds + \int_0^t Z_s dN_s \tag{2.6}$$

where:

$$Z_t = \Delta X_{T_n} \mathbb{1}_{[T_n, T_{N+1})}(t) \tag{2.7}$$

where $\{T_n\}$ are the jump times of X_t.

We now state and prove an important result called the Rate Conservation Law (RCL) first shown by Miyazawa that is very useful in the context of the results in this chapter.

Theorem 2.5   Rate Conservation Law

Let $\{X_t\}$ be a stationary càdlàg process that is consistent with a flow θ_t on $(\Omega, \mathcal{F}, P)$ with $\mathbb{E}|X_t| < \infty$, $t < \infty$.

Then:

$$\mathbb{E}[X_0^+] + \lambda_N \mathbb{E}_N[\Delta X_0] = 0 \tag{2.8}$$

where $\lambda_N = \mathbb{E}N(0,1]$ where $N(.)$ is the stationary point process associated with the jumps and $\mathbb{E}_N[.]$ is the expectation w.r.t. the corresponding Palm measure.

Proof. The proof is straightforward given the definition of the Palm probability, stationarity of $\{X_t\}$ and equation (2.3).

Indeed, taking expectations in (2.3) we obtain:

$$\mathbb{E}[X_t] = \mathbb{E}[X_0] + \mathbb{E}[\int_0^t X_s^+ ds] + \mathbb{E}[\sum_{0 < s \le t} \Delta X_s]$$

Noting that $\mathbb{E}[X_t] = \mathbb{E}[X_0]$ and $\mathbb{E}[\sum_{0 < s \le t} \Delta X_s] = \mathbb{E}[\int_0^t \Delta X_s dN_s] = t\lambda_N \mathbb{E}_N[\Delta X_0]$ by the definition of the Palm probability, the result follows. $\qquad\square$

Remark 2.6 The RCL is easily extended to continuous and differentiable functions which we state below.

Let $f(.) : \Re \to \Re$ be C^1 with $\mathbb{E}[f(X_0)] < \infty$

Then:

$$\boxed{\mathbb{E}[f'(X_0)X_0^+] + \lambda_N \mathbb{E}_N[\Delta f(X_0)] = 0} \qquad (2.9)$$

where $f'(x) = \frac{df(x)}{dx}$ and $\Delta f(X_0) = f(X_0) - f(X_{0-})$.

Another related and more general formula is one that is referred to as the *Swiss Army formula,* due to Brémaud, so-called because of its versatility. Essentially this follows from the Lebesgue-Stiltjes integration formula. In deriving the RCL the continuous and jump terms are associated with the process $\{X_t\}$. The Swiss Army formula allows us to generalize the result to jointly defined processes and thus the jumps and the continuous parts can correspond to different processes (analogous to choosing different blades of a Swiss army knife) that are related by a flow equation. We can show that the RCL is a particular case of the Swiss Army formula by defining the appropriate processes. We state and prove the result below. It follows directly from the Lebesgue-Stieltjes integration by parts formula.

Theorem 2.7 Swiss Army Formula of Palm Calculus

Let $\{A(t)\}$ be a stationary point process whose points are denoted by $\{T_n\}$ with $\ldots < T_{-1} < T_0 \le 0 < T_1 < \ldots$. Let $\{W_n\}$ be a stationary sequence of non-negative random variables and define $\tau_n = T_n + W_n$.[2] Let $D(t)$ be the point process associated with the sequence $\{\tau_n\}$ i.e. $D(t) = \sum_{n \in Z} \mathbf{1}_{(0,t]}(\tau_n)$. Let $X(t) = X(0) + A(t) - D(t)$, $X(0)$ a positive integer, then $\{X(t)\}$ is an integer-valued càdlàg

---

[2]Note the sequence $\{\tau_n\}$ need not be ordered.

process. Let $B(t)$ be a stationary non-decreasing càdlàg process. Then for any stationary process $Z(t)$ that is jointly stationary with $A(.)$ and $B(.)$:

$$\lambda_A \mathbb{E}_A[\int_{(0,W_0]} Z(s)dB(s)] = \frac{1}{t}\mathbb{E}[\int_{(0,t]} X(s-)Z(s)dB(s)] \tag{2.10}$$

Proof. By definition $X(t) - X(0) = A(t) - D(t)$. Note (2.4) can be re-written as:

$$X_t Y_t = X_0 Y_0 + \int_{(0,t]} X_{s-}dYs + \int_{(0,t]} Y_s dX_s$$

Take as $Y_t = \int_{(0,t]} Z(s)dB(s)$, $X_t = X(t) - X(0)$ and noting $Y_0 = 0$ substituting in (2.4) we obtain:

$$(X(t) - X(0))\int_{[(0,t]} Z(s)dB(s) = \int_{(0,t]} (X(s-) - X(0))Z(s)dB(s) + \int_{(0,t]}\int_{(0,s]} Z(u)dB(u)dX(s)$$
$$\tag{2.11}$$

Noting that $(X(t) - X(0))\int_{[(0,t]} Z(s)dB(s) = \int_{(0,t]}\int_{(0,t]} Z(s)dB(s)dX(u)$ after re-arrangement (2.11) can be written as:

$$\int_{(0,t]}\left(\int_{[s,t]} Z(u)dB(u)\right)dX(s) + X(0)\int_{(0,t]} Z(s)dB(s) = \int_{(s,t]} X(u-)Z(u)dB(u) \tag{2.12}$$

Now noting that $dX(t) = dA(t) - dD(t)$ we obtain that the first term on the lhs above:

$$\int_{(0,t]}\int_{(s,t]} Z(u)dB(u)dX(s) = \sum_n \int_{(T_n,t]} Z(u)dB(u)\mathbb{1}_{(0,t]}(T_n) - \sum_n \int_{(\tau_n,t]} Z(u)dB(u)\mathbb{1}_{(0,t]}(\tau_n)$$
$$\tag{2.13}$$

Define $N(t) = \{n : T_n \leq t, \tau_n > t\}$. By definition $X(t) = card(N(t))$. Noting that the second term of (2.12) can be written as: $\sum_{n \in N(0)}\int_{(0,t]} Z(s)dB(s) = \sum_{n \in N(0)}(\int_{(0,\tau_n]} Z(s)dB(s) + \int_{(\tau_n,t]} Z(s)dB(s)$. Furthermore,

$$\sum_n \int_{(\tau_n,t]} Z(u)dB(u)\mathbb{1}_{(0,t]}(\tau_n) = \sum_{n \in N(0)} \int_{(\tau_n,t]} Z(s)dB(s)\mathbb{1}_{[T_n \leq 0]}\mathbb{1}_{(0,t]}(\tau_n)$$
$$+ \sum_n \int_{(\tau_n,t]} Z(s)dB(s)\mathbb{1}_{(0,t]}(T_n)\mathbb{1}_{(0,t]}(\tau_n)$$

We can thus re-group terms using the above and (2.13) to show that the lhs of (2.12) is given by:

$$\int_{(0,t]}\left(\int_{[s,t]} z(u)db(u)\right)dX(s) + X(0)\int_{(0,t]} Z(s)dB(s) = \sum_n \int_{(T_n,\tau_n]} Z(u)dB(u)\mathbb{1}_{(0,t]}(T_n)$$
$$+ R(0) - R(t) \tag{2.14}$$

where:

$$R(t) = \sum_{n \in N(t)} \int_{(t, \tau_n)} Z(u) dB(u)$$

From stationarity $R(t) = R(0) \circ \theta_t$ and so $\mathbb{E}[R(t)] = \mathbb{E}[R(0)]$. Hence, taking expectations on both sides of (2.12), using (2.14) to represent the lhs, and the definition of Palm probability associated with $A(t)$ we obtain that:

$$t \lambda_A \mathbb{E}_A[\int_0^{W_0} Z(u) dB(u)] = \mathbb{E}[\int_0^t X(s-)Z(s) dB(s)] .$$

Noting that $\mathbb{E}[A(0, t)] = \lambda_A t$, from which the result follows. $\qquad\square$

Both the RCL and the Swiss Army formula are very versatile. Indeed, the RCL can be viewed as a special case of the Swiss Army formula but we will still use both forms here. The RCL allows us to derive many important queueing formulae. Before doing so let us see three simple applications of the RCL: 1) derive the Palm inversion formula, 2) proof of Neveu's exchange formula, and 3) obtain the Brill and Posner level crossing formula. We will provide applications of (2.10) later when we study queues.

2.1.1 APPLICATIONS OF THE RCL AND SWISS ARMY FORMULA

1. Derivation of the Palm Inversion Formula

Define $T_+(t) = \inf_n \{T_n : T_n > t\}$ and $Y_t = \int_t^{T_+(t)} f(X_s) ds$. Then by direct calculation: $Y_t^+ = -f(X_t)$, and noting that $T_+(T_m) = T_{m+1}$ and $T_+(T_m-) = T_m$ by definition of $T_+(t)$, we obtain:

$$
\begin{aligned}
\Delta Y_{T_m} &= \int_{T_m}^{T_+(T_m)} f(X_s) ds - \int_{T_m-}^{T_+(T_m-)} f(X_s ds) \\
&= \int_{T_m}^{T_{m+1}} f(X_s) ds - \int_{T_m-}^{T_m} f(X_s) ds \\
&= \int_{T_m}^{T_{m+1}} f(X_s) ds
\end{aligned}
$$

Now applying the RCL and noting that $\mathbb{P}_N \circ \theta_{T_m} = \mathbb{P}_N$ and $\int_{T_m}^{T_{m+1}} f(X_s) ds = \theta_{T_m} \circ \left(\int_0^{T_1} f(X_s) ds \right)$ it follows that:

$$\mathbb{E}[f(X_0)] = \lambda_N \mathbb{E}_N[\int_0^{T_1} f(X_s) ds]$$

The above result can also be directly from the Swiss Army formula by taking $W_n = T_{n+1} - T_n$, $dB(s) = ds$, and noting by definition $X(t) = 1$ for this choice of W_n.

2. Proof of Neveu's exchange formula

For convenience let us call $N = N^1$ and $N' = N^2$. Let T_i^1 and T_i^2 be the points of N^1 and N^2 respectively. Let $N = N^1 + N^2$. Let $T_-^2(t)$ be the last point of $N^2 \leq t$. Define:

$$Y_t = \int_{(T_-^2(t),t]} X_s dN_s^1$$

Then, it is easy to see $Y_0^+ = 0$ $\mathbb{P}$ a.s. w.r.t. dt since the integrand is singular w.r.t. Lebesgue measure—it is defined only at jumps of N^1.

Let $\mathbb{1}_{[0 \in N]}$ denote the indicator that 0 is a point of N, i.e., $\mathbb{1}_{[\Delta N_0 = 1]}$. Then under $\mathbb{P}_N$,

$$Y_0 = X_0 \mathbb{1}_{[0 \in N^1]} + \left(\int_{(T_-^2,0)} X_s dN_s^1 \right) \mathbb{1}_{[0 \in N^1]} + \left(\int_{(T_-^2(0),0]} X_s dN_s^1 \right) \mathbb{1}_{[0 \in N^2]}$$

and by definition

$$Y_{0-} = \int_{(T_-^2(0-),0-]} X_s dN_s^1 = \int_{(T_{-1}^2(0),0)} X_s dN_s^1$$

Note $T_-^2(0-) = T_{-1}^2(0)$ by convention.

Then applying the RCL to $\{Y_t\}$ we obtain:

$$\mathbb{E}_N[Y_0 - Y_{0-}] = 0$$

Noting that $\mathbb{E}_N[X \mathbb{1}_{[0 \in N^i]}] = \mathbb{E}_{N^i}[X]\mathbb{P}(0 \in N^i)$, $\mathbb{P}_N(0 \in N^1) = \frac{\lambda_1}{\lambda_1 + \lambda_2} = 1 - \mathbb{P}_N(0 \in N^2)$, and the fact that under $\mathbb{P}_{N^2}$ we have $T_-^2(0) = 0$ we obtain

$$\mathbb{E}_N[Y_0] = \mathbb{E}_{N^1}[X_0]\mathbb{P}_{N^1}(0 \in N^1) + \mathbb{E}_{N^1}[\int_{(T_{-1}^2(0),0)} X_s dN_s^1]\mathbb{P}_N(0 \in N^1)$$

and

$$\mathbb{E}_N[Y_{0-}] = \mathbb{E}_{N^1}[\int_{(T_{-1}^2(0),0)} X_s dN_s^1]\mathbb{P}_N(0 \in N^1) + \mathbb{E}_{N^2}[\int_{(T_{-1}^2,0)} X_s dN_s^1]\mathbb{P}_N(0 \in N^2)$$

Hence, we obtain:

$$\mathbb{E}_{N^1}[X_0]\mathbb{P}_N(0 \in N^1) = \mathbb{E}_{N^2}[\int_{(T_{-1}^2,0)} X_s dN_s^1]\mathbb{P}_N(0 \in N^2)$$

from which we obtain Neveu's exchange formula

$$\lambda_1 \mathbb{E}_{N^1}[X_0] = \lambda_2 \mathbb{E}_{N^2}[\int_0^{T_1^2} X_s dN_s^1]$$

where we have used the fact that $\mathbb{P}_{N^2} = \mathbb{P}_{N^2} \circ \theta_{0-T_{-1}^2}$ by the shift invariance of P_{N^2} w.r.t $\{T_i^2\}$.

To obtain this from the Swiss Army formula, let $B(t)$ be a stationary point process N_2 with mean intensity λ_2. Take $W_n = T_{n+1}^1 - T_n^1$ where T_n^1 are the points of $N_1(t) = A(t)$. Then noting by definition of Palm probability that $\mathbb{E}[\int_{(0,1]} Z(s)dN_s^2] = \lambda_2\mathbb{E}_{N^2}[Z(0)]$ we obtain Neveu's exchange formula.

3. Level Crossing Formula We can use the RCL to obtain a very useful result that relates the probability density of a process (if it exists) to level crossings of the process that leads to the so-called level crossing formula due to Brill and Posner.

Proposition 2.8   *Let $\{X_t\}$ be a stationary càdlàg process that possesses a density denoted by $p(x)$. Then:*

$$p(x)E[X_0^+|X_0 = x] = \lambda_N E_N[\mathbf{1}_{[X_{0-}>x]}\mathbf{1}_{[X_0\leq x]} - \mathbf{1}_{[X_{0-}\leq x]}\mathbf{1}_{[X_0>x]}] \tag{2.15}$$

Proof. The proof follows by directly applying the RCL to the process: $Y_t = \mathbf{1}_{[X_t>x]}$. Formally (since the indicator function is not differentiable), $Y_t^+ = \delta(X_t - x)X_t^+$ and hence, $\mathbb{E}[Y_0^+] = p(x)\mathbb{E}[X_0^+|X_0 = x]$. Substituting in the RCL we obtain:

$$p(x)\mathbb{E}[X_0^+|X_0 = x] = \lambda_N\mathbb{E}_N[\mathbf{1}_{[X_{0-}>x]} - \mathbf{1}_{[X_0>x]}]$$

Noting that:

$$\mathbf{1}_{[X_{0-}>x]} - \mathbf{1}_{[X_0>x]} = \mathbf{1}_{[X_{0-}>x]}\mathbf{1}_{[X_0\leq x]} - \mathbf{1}_{[X_{0-}\leq x]}\mathbf{1}_{[X_0>x]}$$

We obtain the result as stated.   □

The interpretation of the result is that the probability density at x multiplied by the derivative of the process at x is equal to the rate of the down crossings of x minus the rate of the up crossings of x.

In the sequel we will focus on queueing models and obtain some important formulae of relevance to performance evaluation via the application of the RCL and the integration by parts formula for càdlàg processes.

2.2   QUEUEING MODELS

Queues can be mathematically modeled in a number of ways. Two of the most common ways of looking at queues are: 1) from a customer or packet viewpoint, and 2) from the viewpoint of the amount of work in the system. The first characterization leads to a model where the number of customers or packets[3] in the queue[4] at a given time is studied that leads to a state space that is at least partly discrete but that might not be sufficient for a complete mathematical (probabilistic) description of the queue behavior. The second characterization is in terms of the work in the system

---

[3]We will refer to this as the packet level modeling in keeping with networking applications
[4]Often referred to as the congestion process, a term employed in this monograph.

and is often referred to as the *unfinished work* or *virtual waiting time* or *workload* in the queue. We will use the term workload in these notes. The aim is to write down a mathematical evolution of the process and study properties of the solution from both a trajectorial or sample-path view point as well as a probabilistic viewpoint.

In order to specify a queueing model we need to specify a number of quantities. These are:

1. The arrival process to the queue. More precisely, we need to specify the probabilistic structure of the arrival process, i.e., whether it is a point process or a fluid arrival process; how much work an arrival brings to the system, for example in the context of networks the packet length and whether this is fixed or variable, etc.

2. The mechanism by which the server processes the arriving work, i.e., whether the system is work conserving or not, the server processes packets according to the order of their arrival (First In First Out (FIFO)), or whether the server processes shorter packets first, etc.

3. The number of servers available.

4. Whether there is a buffer or not and the size of the buffer.

In the models we consider we will assume that the system is *work conserving*, i.e., the server cannot be idle if there is work in the system.

A compact notation to denote queues of the form $A/B/C/K/m/Z$ was introduced by Kendall. The various symbols are:

1. A denotes the arrival process. If M is used it signifies that the input is a Poisson process and thus we need to specify the intensity λ. If the first symbol is GI it signifies that the input is renewal and we need to specify the distribution $F(t)$ of the inter-arrival times. If the first symbol is G it specifies that the input is a stationary point process.

2. B denotes the distribution of the service time. Most switches have servers that work at a constant rate in which case the service time distribution is the packet length distribution. If it is M it implies that the service times are exponentially distributed. If it is GI the service times are i.i.d. with distribution $G(t)$. If it is G then the service times are identically distributed with distribution $G(t)$ but need not be independent. Many books do not differentiate between GI and G.

3. C denotes the number of servers. Typically, $C = 1$ or some integer and could be infinite.

4. K denotes the total number of places in the system. If $K = C$ then there is only room in the server and none for additional packets. In most cases we take $K = \infty$ implying unlimited buffer space.

5. m is used to denote the population of users. In most cases $m = \infty$.

6. Z usually is used to denote whether the service discipline is FIFO (First In First Out also referred to as First Come First Served (FCFS)), LIFO (Last in First Out), SPT (Shortest Processing Time First), Random (packets are processed in random order), etc. We will mostly be concerned with FIFO queues and thus this information is not appended in the notation.

In most commonly used models the last three elements of Kendall's notation are not specified and in that case the assumption is $K = \infty, m = \infty$, and $Z = FIFO$. In these notes we will focus on $M/M/C/K$, $M/G/1$, $M/G/\infty$, $GI/M/1$, and fluid queues. Throughout we will assume that the service discipline is FIFO though many of the results we will see only require the work conserving assumption. These models are the most tractable and yield many useful insights. Moreover, because of the superposition theorem for independent point processes given in Chapter 1, the Poisson assumption for arrivals can often be justified.

We now develop the evolution equations for queues from the two standpoints that we mentioned earlier namely from the point of view of occupancy in terms of the number of packets and in terms of the workload process in a queue.

The typical sample-paths or trajectories of the queueing process are indicated in the figure 2.1 below.

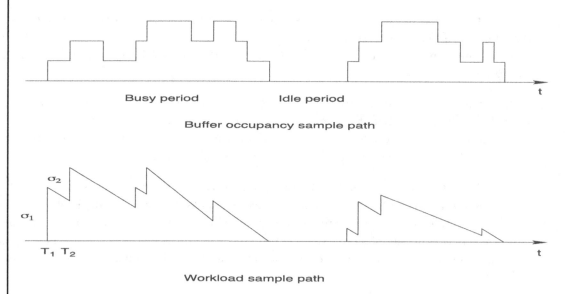

Figure 2.1: Typical sample-paths of queues viewed from a congestion process and the workload viewpoints.

We begin by first viewing the queueing process in terms of the buffer occupancy or congestion process. Here the sample-path of the queue increases by +1 whenever there is an arrival to the queue at times $\{T_n\}$ that denote the arrival instants and decreases by 1 whenever packets depart from the queue at instants that are denoted by $\{d_n\}$.

2.2.1 QUEUES VIEWED FROM CONGESTION PROCESS VIEWPOINT

Let $A(s, t]$ denote the number of arrivals that occur and $D(s, t]$ denote the number of departures in the interval $(s, t]$. We assume throughout that the queue is work conserving in that the server does not stop when there are packets in the queue. We also assume that the service discipline is FIFO.

Let Q_t denote the number of packets in the queue at time t or the congestion process at time t. Then:

$$Q_t = Q_0 + A(0, t] - D(0, t] \quad t \geq 0 \tag{2.16}$$

Based on the evolution equation above it is clear that Q_t is a discrete valued process taking values in $\mathbb{Z}_+$ (the non-negative integers) and the sample paths are constant between jumps (of value ± 1) arrivals and departures. By construction the process is right continuous. Let us make some simple observation: If $\liminf_{t \to \infty} \frac{A(0,t] - D(0,t]}{t} > 0 \; a.s.$ then $Q_t \to \infty \; a.s..$ Such a situation implies that the queue is unstable. Our interest is when the queue is finite. This is related to the notion of stability and there are many different notions; for example $\sup_t Q_t < \infty \; a.s.$, $\limsup_{t \to \infty} \frac{1}{t} \int_0^t Q_s ds < \infty$,etc. In these notes our interest is to study conditions when stationary arrival and departure processes result in Q_t being a stationary process and conditions on the processes such that the resulting buffer occupancy has a finite mean or higher moments and try to characterize them explicitly. In the sequel we will use A_t and D_t to denote $A(0, t]$ and $D(0, t]$ respectively.

However, before we state and prove the main stability result let us see some implications of stability in the sense that Q_t is stationary with finite mean (and hence $A(.)$ and $D(.)$ are stationary increment processes). Let us assume that Q_t is stationary and has a finite mean, taking expectations in (2.16), $\mathbb{E}[Q_t] = \mathbb{E}[Q_0]$ and hence $\mathbb{E}[A(0, t]] = \mathbb{E}[D(0, t]]$ for all t implying that if $\lambda_A = \mathbb{E}[A(0, 1]]$ denotes the mean arrival rate then: $\lambda_A = \lambda_D \stackrel{d}{=} \mathbb{E}[D(0, 1]]$ the mean departure rate, in other words the average input rate is equal to the average output rate.

Let us see another relation that follows from the queue evolution, the Lebesgue-Stieltjes integration formula and the definition of palm probabilities. First note that since $Q_t^+ = 0$ a.e. t. (the trajectories are piecewise constant except at jumps), for any measurable, function $f(.)$:

$$f(Q_t) = f(Q_0) + \int_0^t [f(Q_s) - f(Q_{s-})] dA_s + \int_0^t [f(Q_s) - f(Q_{s-})] dD_s \tag{2.17}$$

Noting that at jumps T_n of A_t we have $f(Q_{T_n}) = f(Q_{T_{n-}} + 1)$ and at jumps d_n of D_t, $f(Q_{d_n}) = f(Q_{d_{n-}} - 1) \mathbb{1}_{[Q_{d_{n-}} > 0]}$ since departures can only take place when there are packets in the queue.

Now choose $f(x) = \mathbb{1}_{[x=n]}$ and if $\{Q_t\}$ is stationary, substituting in (2.17), noting $\mathbb{E}[\mathbb{1}_{[Q_t=n]}] = \mathbb{E}[\mathbb{1}_{[Q_0=n]}]$ we obtain, for $n \geq 1$:

$$\mathbb{E}[\int_0^t (\mathbb{1}_{[Q_{s-}+1=n]} - \mathbb{1}_{[Q_{s-}=n]}) dA_s] = \mathbb{E}[\int_0^t (\mathbb{1}_{[Q_{s-}-1=n]} - \mathbb{1}_{[Q_{s-}=n]}) dD_s]$$

Define $\mathbb{P}_A(Q_{0-} = n) = \pi_A(n)$ and $\mathbb{P}_D(Q_{0-} = n) = \pi_D(n)$, then by the definition of the Palm probability we obtain:

$$\lambda_A[\pi_A(n-1) - \pi_A(n)] = \lambda_D[\pi_D(n+1) - \pi_D(n)]$$

But we have seen $\lambda_A = \lambda_D$ and furthermore by considering the case $n = 0$ and noting that the state space (or values that Q_t can take) is in $\{0, 1, 2, \cdots\}$ and there are no departures from an empty queue, we see: $\pi_A(0) = \pi_D(1)$ and substituting in the equation above we obtain:

$$\pi_A(n) = \pi_D(n+1), \quad n = 0, 1, 2, \ldots \tag{2.18}$$

In other words, we have shown that the distribution that an arrival *sees* is equal to the stationary distribution that a departure *leaves*. Of particular interest is the following important observation in the $M/G/1$ case when the arrivals are Poisson. In this case, from PASTA we know that $\pi_A(n) = \pi(n)$ where $\pi(n)$ is the (time)[5] stationary distribution or the stationary distribution under $\mathbb{P}$ and so if we view the queue at departure times and compute its stationary distribution we obtain the time-stationary distribution of the queue. This important observation is the basis of calculations for $M/G/1$ queues made by considering the queueing process only at departure times.

We state these results as a proposition.

Proposition 2.9

Let Q_t be the congestion process resulting from a stationary arrival process A_t and a corresponding departure process D_t. If Q_t is stationary, then the stationary distribution seen just before an arrival when the buffer is in state n is equal to the stationary distribution just after a departure in state $n + 1$ i.e.,

$$\boxed{\pi_A(n) = \pi_{D+}(n), \quad n = 0, 1, 2, \ldots} \tag{2.19}$$

where $\pi_A(.)$ denotes the distribution just before an arrival, and $\pi_{D+}(.)$ denotes the distribution just after the departure.

Moreover, if the queue is of the $M/G/1$ type then:

$$\pi_{D_+}(n) = \pi(n), \quad n = 0, 1, 2, \ldots$$

where $\pi(n)$ is the stationary distribution of the queue.

In the general case, it is of interest to compute quantities under the stationary distribution. For stationary (and ergodic) single server queues there is in very simple relationship that can be found in terms of so-called conditional intensities that we develop below.

Define:

$$\begin{aligned}
\lambda_i &= \lim_{t \to \infty} \frac{\int_0^t \mathbb{1}_{[Q_{s-}=i]} dA_s}{\int_0^t \mathbb{1}_{[Q_s=i]} ds} \\
&= \lim_{t \to \infty} \frac{\frac{A_t}{t} \frac{1}{A_t} \int_0^t \mathbb{1}_{[Q_{s-}=i]} dA_s}{\frac{1}{t} \int_0^t \mathbb{1}_{[Q_s=i]} ds}
\end{aligned} \tag{2.20}$$

[5]We will refer to the time stationary distribution simple as the stationary distribution

Note the first term in the numerator above $\frac{A_t}{t} \to \lambda_A$, the second term in the numerator goes to $\pi_A(i)$ while the denominator goes to $\pi(i)$ as $t \to \infty$.

Similarly, define

$$\mu_i = \lim_{t \to \infty} \frac{\int_0^t \mathbb{1}_{[Q_{s-}=i]} d D_s}{\int_0^t \mathbb{1}_{[Q_s=i]} ds} \tag{2.21}$$

Therefore, by the definition of λ_i and μ_i we obtain: $\lambda_i \pi_i = \lambda_A \pi_A(i)$ and $\mu_i \pi(i) = \lambda_D \pi_D(i)$, then in light of (2.19) above we obtain the following relationship:

$$\lambda_i \pi(i) = \mu_{i+1} \pi(i+1), \quad i = 0, 1, 2, \ldots \tag{2.22}$$

and hence:

$$\pi(n) = \prod_{i=1}^n \frac{\lambda_{i-1}}{\mu_i} \pi(0), \quad n = 1, 2, \ldots \tag{2.23}$$

Remark 2.10 This is called the first-order equivalent of a queue by Brémaud for the following reason: the form of the stationary distribution is exactly the same as a birth-death process (see Appendix) with birth rates λ_i in state i and death rates μ_i in state i. Equation (2.22) is called the detailed balance equation. From the fact that birth-death processes are reversible in equilibrium (stationary state) it can be shown that if a birth death process is restricted to a closed subset $S \subset E$ of its state space E then for any $i \in S$, the stationary probability distribution of the restricted process is given by:

$$\pi_S(i) = \frac{\pi(i)}{\sum_{i \in S} \pi(i)} \quad i \in S$$

which is a useful result to keep in mind in the sequel.

Of course in the general case the difficulty is to find λ_i and μ_i for general arrival processes and departure processes. A special case when we can determine these rates is the $M/M/1$ case. In this case, the arrival process is Poisson with rate λ and the service rate is exponential with mean $\frac{1}{\mu}$. Because of the memoryless property of the exponential distribution the departure rate is μ. We state this result below:

Proposition 2.11 *Let Q_t be the congestion process of a $M/M/1$ queue in equilibrium. Then the stationary distribution is given by:*

$$\pi(n) = \rho^n (1 - \rho), \quad n = 0, 1, 2, \ldots \tag{2.24}$$

where $\rho = \frac{\lambda}{\mu} < 1$ is called the traffic intensity.

Proof. It follows directly from (2.23) and noting that $\lambda_i = \lambda \; \forall i$ and $\mu_i = \mu \; \forall i$ and $\pi(0) = 1 - \rho > 0$ can be computed from the fact that $\sum_{i=0}^{\infty} \pi(i) = 1$. Condition $\rho < 1$ is necessary and sufficient for Q_t to possess a stationary distribution (and be ergodic). $\qquad \square$

An immediate corollary is:

Corollary 2.12 *If Q_t is the congestion process of a $M/M/C$ queue then:*

$$
\begin{aligned}
\pi(n) &= \frac{\rho^n}{n!}\pi(0), \quad n = 0, 1, 2, \ldots, C \\
&= \frac{\rho^n}{C!C^{n-C}}\pi(0), \quad n = C+1, C+2, \ldots,
\end{aligned}
$$

and $\pi(0)$ is determined from $\sum_{i=0}^{\infty} \pi(i) = 1$, $\rho = \frac{\lambda}{\mu}$ and $\rho < C$.

Proof. In this case, it is sufficient to note that if there are C servers then the departure rate from state n is $\mu_n = n\mu$, $n = 1, 2, \ldots, C$ and $\mu_n = \mu C$, $n > C$ since it is the rate corresponding to the minimum of n exponential r.v.'s each of rate μ. The rest follows from equation (2.23). □

The $M/M/C/K$ (here $K \geq C$) case can easily be handled in light of the remarks above by noting that in this case:

$$
\pi_K(i) = \frac{\pi(i)}{\sum_{i=0}^{K} \pi(i)}
$$

where $\pi(i)$ is the stationary distribution corresponding to the $M/M/C$ queue in equilibrium. In this case, since $K < \infty$ we do not need $\rho < C$ and there is no need to evaluate $\pi(0)$ because it cancels in the numerator and denominator.

The infinite server queue has special properties that we will discuss in detail later. For the time being we leave our discussion of queues viewed from a congestion viewpoint and now consider the second approach of viewing queues from the viewpoint of workloads.

2.2.2 QUEUES FROM A WORKLOAD VIEWPOINT

Queues viewed from a workload point of view have sample-paths as shown in Fig. 2.1. As indicated, the sample-paths have jumps corresponding to arrivals and the size of the jumps, denoted by the sequence $\{\sigma_n\}$, corresponds to the amount of work brought to the system by an arrival. In between jumps the workload decreases at a rate determined by the server speed which is assumed to be a constant and denoted by c. Let $\{W_t\}$ denote the workload at time t. It denotes the amount of work to be processed at time t in the queue. Clearly, the sample-path of W_t is càdlàg and the jumps correspond to σ_n. The evolution of the workload can be written as:

$$
W_t = W_0 + X(0, t] - c \int_0^t \mathbb{1}_{[W_s > 0]} ds, \quad t \geq 0 \tag{2.25}
$$

where $X(0, t]$ denotes the amount of work arriving in $(0, t]$ and is given by $X(0, t] = \sum_{0 < T_n \leq t} \sigma_{T_n} = \int_0^t \sigma_s dA_s$ where $\sigma_t = \sigma_{T_n} \mathbb{1}_{[T_n, T_{n+1})}(t)$. Note we can re-write equation (2.25) as:

$$
W_t = W_0 + X(0, t] - ct + c \int_0^t \mathbb{1}_{[W_s = 0]} ds = W_0 + J(0, t] + I(0, t] \tag{2.26}
$$

where $J(0, t] = X(0, t] - ct$ is called the *netput* and $I(0, t] = c \int_0^t \mathbb{1}_{[W_s=0]} ds$ is called the idle time. This form of writing the workload process has a very nice interpretation. It is called a Skorokhod reflection problem stated as follows.

Given a càdlàg process x_t find a non-negative process z_t such that

$$y_t = x_t + z_t \geq 0 \; t \geq 0$$

Once again denoting $J(0, t]$ and $I(0, t]$ by J_t and I_t respectively, we note the following.

Remark 2.13

1. I_t is an increasing process in t with $I_0 = 0$.

2. $\int_0^t W_s dI_s = 0$, i.e., I_t can only increase when W_t is 0.

3. Furthermore, $I_t = \sup_{0<s\leq t}(-J_s - W_0)^+$ where $(x)^+ = \max\{x, 0\}$ This last result needs proof. Without loss of generality let us take $W_0 = 0$ because we can include it in J_t by setting $J_0 = W_0$.

   Let us show that with I_t as defined we have $\int_0^t W_s dI_s = 0$. Indeed, suppose $dI_t > 0$ this implies that the supremum should be attained at t or in other words, $I_t = -J_t$ and then $W_t = 0$. So at any point of increase of I_t we see $W_t = 0$ and thus I_t as defined satisfies properties 1 and 2 above.

   I_t is also called the regulator process in the context of the Skorohod reflection problem. It is the amount we need to *push* W_t so that it remains non-negative.

From the characterization of the idle time process I_t we immediately have:

$$\boxed{W_t = J_t + I_t = J_t - \inf_{0<s\leq t}(J_s)^+ = \sup_{0<s\leq t}(J_t - J_s)^+ = \sup_{0<s\leq t}(J(s, t])^+} \quad (2.27)$$

The above equation is sometimes called Lindley's equation. Since the r.h.s. is completely specified by the process $J(s, t]$ and the operation $(x)^+$ is continuous, the writing of W_t as $\psi(J(., t])$ is for an appropriately defined function $\psi(.)$, here $\sup_{0<s\leq t}(J(s, t])^+$, called the reflection map.

Let us re-write things slightly differently, instead of calculating the workload at t starting from 0 assume we go backwards in time, then, assuming that $W_{-\infty} = 0$ we obtain:

$$\boxed{W_0 = \sup_{t\geq 0}(X(-t, 0] - ct)^+} \quad (2.28)$$

The workload formulation is particularly convenient to study the stability and ergodicity issue that we have assumed so far. Moreover, although we have assumed that the input process is a point process this has not been used so far. All the results obtained hold true for any increasing process $X(0, t]$. We will exploit this fact when we study fluid queues at the end of the chapter. But before doing so let us see some general properties about the workload process. Suppose that $\limsup_{t \to \infty} \frac{X(-t, 0]}{t} < c$ $a.s.$ then we see that $X(-t, 0] - ct \to -\infty$ $a.s.$ when $t \to \infty$ and so the supremum in (2.28) must be finite.

Let us now assume that the point process $X(0, t]$ is ergodic and $\lim_{t \to \infty} \frac{X_t}{t} = \rho < c$. Then, since W_t is finite $a.s.$ dividing (2.25) on both sides by t we obtain:

$$\lim_{t \to \infty} \frac{1}{t} \int_0^t \mathbb{1}_{[W_s > 0]} ds = \mathbb{P}(W_0 > 0) = \frac{\rho}{c} = 1 - \lim_{t \to \infty} \frac{1}{t} \int_0^t \mathbb{1}_{[W_s = 0]} ds$$

Now let us study the stability result and prove the main stability result for single-server work conserving queues when the inputs are stationary point processes. Let $\{T_n\}$ denote the points of the arrival process A_t and let the sequence $\{\sigma_n\}$ denote the work brought in by the arrivals where σ_n denotes the work brought by the arrival at T_n. The sequences are assumed to be consistent w.r.t. θ_{T_n}. We assume that $\lambda_A \mathbb{E}_A[\sigma_0] = \rho$ and without loss of generality let us take $c = 1$. The idea of this proof goes back to Loynes and is called a Loynes construction.

Let $W_n = W_{T_{n-}}$, $W_0 = w$, and let $\tau_n = T_{n+1} - T_n$. Then, assuming $W_0 = 0$,

$$W_1 = (\sigma_0 - \tau_0)^+$$

and

$$W_{n+1} = (W_n + \sigma_n - \tau_n)^+$$

Define $Y_n = \sigma_n - \tau_n$ Iterating we obtain:

$$\begin{aligned} W_n &= \max\left(Y_n, Y_n + Y_{n-1}, \ldots, \sum_{k=0}^n Y_k \right)^+ \\ &= \sup_{0 \le m \le n} \left(\sum_m^n Y_k \right)^+ \end{aligned}$$

or equivalently, starting with $W_{-n} = 0$:

$$W_0^n = \sup_{-n < m \le 0} \left(\sum_{k=-m}^0 Y_k \right)^+$$

If $\mathbb{E}[Y_k] < 0$ it implies that the sequence $\sum_{k=-m}^0 Y_k \to -\infty$ $a.s.$ as $m \to \infty$ which implies that $W_0 < \infty$ $a.s.$ Now it is clear that since the supremum over $(-n, 0]$ is less than or equal to the supremum over $(-(n+1), 0]$ it implies that $\{W_0^n\}$ is a non-decreasing, nonnegative sequence that is $a.s.$ finite and hence must have a limit that we denote by W_0^∞ and in particular:

$$W_0^\infty = (W_0^\infty + \sigma - \tau)^+$$

and since τ and σ are θ_{T_n} consistent and invariant we have:

$$W_0^\infty \circ \theta_{T_k} = (W_0^\infty + \sigma - \tau)^+ \circ \theta_{T_k} = \sup_n (\sum_{m=-n}^{0} Y_m \circ \theta_{T_k})^+ = (W_0^\infty + \sigma - \tau)$$

by the θ_{T_n} invariance of Y_m showing that W_n^∞ is a stationary sequence.

Now if $W_{-n} = w > 0$ there exists $-\infty < N' < -n$ such that $W_{N'} = 0$ since $\sum_{k=-n}^{0} Y_k \to -\infty$ *a.s.* as $n \to \infty$ when $\mathbb{E}[Y_n] < 0$. Thus, we could have started at N' and repeated the same argument with a 0 initial condition.

On the other hand, if $\mathbb{E}[\sigma_n - \tau_n] = \mathbb{E}[Y_n] > 0$ it implies that $\lim_{n \to \infty} \frac{1}{n} \sum_{m=-n}^{0} Y_m > 0$ and hence $W_0 = \infty$ implying the queue is unstable and cannot possess a stationary distribution.

Note that for any $t \in (T_n, T_{n+1}]$

$$W_t = \max(W_n + \sigma_n - (t - T_n))^+$$

and thus from the stationarity w.r.t. θ_{T_n} of W_n we can show that W_t is consistent w.r.t θ_t.

We state this result below .

Theorem 2.14   *Consider a single server work conserving queue that is fed with a stationary sequence (T_n, σ_n) that has a constant server speed of 1. Then, if $\lambda_A \mathbb{E}_A[\sigma_0] < 1$ the queue is stable, i.e., $\sup_t W_t < \infty$ a.s. and moreover the stationary version of the workload is given by:*

$$\boxed{W_0 = \sup_{t \geq 0}(X(-t, 0] - t)^+} \tag{2.29}$$

where $X(0, t] = \int_0^t \sigma_s d A_s$ and $\sigma_s = \sigma_{T_n} \mathbb{1}_{[T_n, T_{n+1}]}(s)$.

Remark 2.15

1. The condition $\mathbb{E}[\sigma - \tau]$ is to be interpreted as $\mathbb{E}_A[\sigma_0] - (\lambda_A)^{-1} < 0$ which is equivalent to $\lambda_A \mathbb{E}_A[\sigma_0] = \rho < 1$

2. The proof for queues with multiple servers is more complicated. The idea of a Loynes type construction in this case is due to Kiefer and Wolfowitz. An interested reader is referred to the book of Baccelli and Brémaud listed in the references. In the multi-server case the condition for stability is $\lambda_A \mathbb{E}[\sigma_0] < C$ where C is the number of servers, each operating at unit rate.

In the next section we will obtain general results on mean waiting times in queues and show an important formula that relates the means of the congestion process and the workload that is the celebrated Little's formula.

2.3   WAITING TIMES AND WORKLOAD IN QUEUES

We now obtain several important results that are useful in determining the performance of queueing systems.

The first important result is a formula that links the average number of packets in the queue to the waiting time of packets in a queue. This allows us to relate the two viewpoints: the congestion viewpoint and the workload viewpoint and is known as Little's formula.

The waiting time of a packet in a queue is the amount of work that the packet sees when it arrives in the queue if the service discipline is FIFO. The sojourn time in the system (also referred to as the delay) is the waiting time + the service time of the packet. There are many approaches to obtain Little's formula and the approach we will follow is one based on the RCL - because the RCL only requires stationarity of a càdlàg process without imposing too many extra hypotheses.

Theorem 2.16 Little's Formula

Let Q_t be the congestion process and W_t be the workload in a stationary single-server, work conserving queue with arrival process $X(0, t]$ as defined in Theorem 2.14 with $\lambda_A \mathbb{E}_A[\sigma] = \rho < 1$ and $\mathbb{E}[W_0] < \infty$.

Then:

$$\boxed{\mathbb{E}[Q_0] = \lambda_A \mathbb{E}_A[W_0]} \tag{2.30}$$

where $W_0 = W_{0-} + \sigma_0$ where W_{0-} denotes the workload seen by an arrival.

Proof. Let us define the following process, referred to as the total sojourn time process (it measures the total remaining work due all the packets in the queue at time t.):

$$V_t = \int_0^t (W_{s-} + \sigma_s - (t-s))^+ dA_s = \int_0^t (W_s - (t-s))^+ dA_s$$

so V_t just counts: $\sum_{j=1}^{Q_t} W_j(t)$ where $W_j(t)$ is the remaining sojourn time of the $j'th$ packet in the queue at time t.

Let us apply the RCL to V_t then we see:

$$V_t^+ = -\int_0^t \mathbb{1}_{[(W_s > (t-s)]} dA_s = -\sum_{T_n} \mathbb{1}_{[T_n \leq t < W_{T_n} + T_n]}$$

which is just the negative of the number of arrivals prior to t that are still in the queue and is therefore equal to Q_t.

On the other hand:

$$\begin{aligned}
\Delta V_{T_n} &= V_{T_n} - V_{T_n-} \\
&= \int_0^{T_n} (W_s - (T_n - s))^+ dA_s - \int_0^{T_n-} (W_s - (T_{n-} - s)^+ dA_s \\
&= \int_{T_n-}^{T_n} (W_s - (T_n - s))^+ dA_s = W_{T_n}
\end{aligned}$$

Therefore, by the RCL

$$\mathbb{E}[Q_0] = \lambda_A \mathbb{E}_A[W_0]$$

$\square$

Remark 2.17 Little's formula can be directly obtained from the Swiss Army formula (2.10) by taking $dB(s) = ds$, W_n is the sojourn or waiting time in the system. Then by definition $X(t)$ denotes the number in the system and hence taking $Z(t) \equiv\equiv 1$:

$$\lambda_A \mathbb{E}_A[W_0] = \mathbb{E}[X(0)]$$

We can further extend Little's formula to relate higher order moments (provided they exist) by choosing $Z(t) = X(t)^k, k \geq 1$ to obtain:

$$\lambda_A \mathbb{E}_A[\int_{(0,W_0]} X^k(s)ds] = \mathbb{E}[X(0)^{k+1}] \tag{2.31}$$

In a similar way if we take $B(t) = A(t)$ then we obtain:

$$\mathbb{E}_A[A(0, W_0]] = \mathbb{E}_A[X(0-)] \tag{2.32}$$

or the average number of arrivals during the sojourn of an a customer is equal to the mean number in the queue just before the arrival under the arrival distribution.

The evaluation of the r.h.s. of Little's formula w.r.t. arrival distribution is important to note. In the case of $M/G/1$ queues, by PASTA, $\mathbb{E}_A[W_0] = \mathbb{E}[W_0]$. Let us see an example where the arrival distribution is different from the stationary distribution and computation with respect to the stationary distribution will not give the right answer.

Example of $M/M/1/K$ queues

This is an example we have seen earlier. In particular, the stationary distribution is given by:

$$\pi(i) = \frac{\rho^i(1-\rho)}{(1-\rho^{K+1})}$$

Hence, $\mathbb{E}[Q_0] = \sum_{i=1}^{K} i\pi(i)$. The mean workload under the stationary distribution is given by:

$$\mathbb{E}[W_0] = \sum_{i=0}^{K-1} \frac{(i+1)}{\mu}\pi(i) = \sum_{m=1}^{K} \frac{m}{\mu}\pi(m-1)$$

since arrivals into the queue can only take place in states $\{0, 1, \dots, K-1\}$ and by the residual time formula the mean workload is the mean of $(i+1)$ service times when the arrival sees i packets in the system. The average arrival rate into the queue is $\lambda(1 - \pi(K))$.

Hence, if we apply Little's formula taking for the r.h.s. the mean workload under the stationary distribution we would obtain noting that $\rho\pi(i) = \pi(i+1)$:

$$\lambda(1 - \pi(K)) \sum_{m=1}^{K} \frac{m}{\mu} \pi(m-1) = (1 - \pi(K)) \sum_{m=1}^{k} m\pi(m) = (1 - \pi(K))\mathbb{E}[Q_0]$$

On the other hand, the arrival stationary distribution is obtained by restricting the $M/M/1$ queue to states $\{0, 1, \ldots, K-1\}$ and we have seen that in that case:

$$
\begin{aligned}
\pi_A(i) &= \frac{\pi(i)}{\sum_{i=0}^{K-1} \pi(i)} \\
&= \frac{\pi(i)}{1 - \pi(K)}
\end{aligned}
$$

Now repeating the calculations we see $\mathbb{E}_A[W] = \frac{\mathbb{E}[Q_0]}{\lambda(1-\pi(K))}$ and thus applying Little's law (correctly) we obtain: $\mathbb{E}[Q_0] = \lambda_A \mathbb{E}_A[W_0]$.

We now obtain another result for the mean waiting time in queues (under FIFO), which in the special case of $M/G/1$ queues is called the Pollaczek-Khinchine formula.

Proposition 2.18

Consider a $G/G/1$ queue in the stationary state with $\lambda_A \mathbb{E}_A[\sigma_0] = \rho < 1$. Then the mean workload (or delay under FIFO) is given by:

$$\mathbb{E}[W_0] = \mathbb{E}_A[\sigma_0 W_{0-}] + \frac{\lambda_A}{2}\mathbb{E}_A[\sigma_0^2] \tag{2.33}$$

In particular, for the case of $M/G/1$ queues the formula reduces to :

$$\mathbb{E}[W_0] = \frac{\mathbb{E}[\sigma_0^2]}{2(1-\rho)} = \frac{\rho}{(1-\rho)} \frac{\mathbb{E}[\sigma_0^2]}{2\mathbb{E}[\sigma_0]} \tag{2.34}$$

Proof. Consider the process $Y_t = (W_t)^2$. Then $Y_t^+ = -2W_t \mathbb{1}_{[W_t > 0]} = -2W_t$ and:

$$
\begin{aligned}
\Delta Y_{T_n} &= W_{T_n}^2 - W_{T_{n-}}^2 \\
&= (W_{T_{n-}} + \sigma_n)^2 - W_{T_{n-}}^2 \\
&= 2W_{T_{n-}}\sigma_n + \sigma_n^2
\end{aligned}
$$

Hence, applying the RCL to Y_t we obtain:

$$2\mathbb{E}[W_0] = 2\lambda_A \mathbb{E}_A[\sigma_0 W_{0-}] + \lambda_A \mathbb{E}_A[\sigma_0^2]$$

In the $M/G/1$ case by PASTA, $\mathbb{E}_A[.] = \mathbb{E}[.]$, $\lambda_A = \lambda$, and from the independence of W_{0-} and σ_0 we obtain $\mathbb{E}[W_{0-}\sigma_0] = \mathbb{E}[W_0]\mathbb{E}[\sigma_0]$ which results in:

$$\mathbb{E}[W_0] = \lambda \frac{\mathbb{E}[\sigma_0^2]}{2(1-\rho)} = \frac{\rho}{1-\rho} \frac{\mathbb{E}[\sigma_0^2]}{2\mathbb{E}[\sigma_0]}$$

$\square$

The last expression on the r.h.s. of (2.34) has been written in terms of the residual service time which we derive by an alternative argument below.

Consider an arbitrary arrival to the queue in equilibrium that sees Q_t packets. Of these one will be in service (because of work conservation) and $Q_t - 1$ will be waiting. Thus, the total sojourn time or workload (including the work brought by the arrival) will be: $\sigma_t + \sum_{i=1}^{Q_t-1} \sigma_i + R(t)\mathbb{1}_{[Q_t>0]}$ where $R(t)$ is the residual service time. Using, Wald's identity (see Appendix) along with the expression for the mean residual service time we obtain:

$$\mathbb{E}[W_0] = \mathbb{E}[Q_0]\mathbb{E}[\sigma_0] + \frac{\mathbb{E}[\sigma_0^2]}{2\mathbb{E}[\sigma_0]}\rho$$

where $\rho = \mathbb{P}(Q_0 > 0)$. Using Little's formula, we have $\mathbb{E}[Q_0] = \lambda_A\mathbb{E}[W_0]$. Hence, substituting for $\mathbb{E}[Q_0]$ we obtain:

$$\mathbb{E}[W_0] = \frac{\rho}{1-\rho}\mathbb{E}[R]$$

which is another way of stating the P-K formula for $M/G/1$ queues.

The P-K formula is often written in terms of the coefficient of variation denoted by $C^2 = \frac{var[\sigma_0]}{(\mathbb{E}[\sigma_0])^2}$ as:

$$\mathbb{E}[W_0] = \frac{\rho}{2(1-\rho)}\mathbb{E}[\sigma_0](C^2+1)$$

Knowing $var(X)$ for a deterministic r.v. is 0 (and thus $C^2 = 0$) and the variance of an exponential r.v. is $(\mathbb{E}[X])^2$ giving $C^2 = 1$, we see that the mean waiting time in a $M/G/1$ queue is smallest for the $M/D/1$ case. The mean waiting time is $\frac{\rho}{1-\rho}\frac{1}{\mu}$ in the $M/M/1$ case

We now treat the case of multiple independent inputs and show that a certain conservation law holds for the mean workload (or delay in the case of FIFO)). We obtain this as a simple consequence of the Lebesgue-Stieltjes integration by parts formula.

Let $\{X_i(0,t]\}_{i=1}^M$ denote M independent stationary increment input processes that are consistent w.r.r. a flow θ_t and $X_i(0,t] = \int_0^t \sigma_{i,s} dA_{i,s}$ where $A_{i,.}$ denotes a stationary point process with average intensity λ_i and $\sigma_{i,s} = \sigma_i \mathbb{1}_{[T_{i,n},T_{i,n+1})}(s)$ where $\{\sigma_{i,n}\}$ are stationary sequences. Suppose $\sum_{i=1}^M \lambda_i \mathbb{E}_{A_i}[\sigma_{i,0}] < 1$. The evolution of the workload due to the aggregate input is given by:

$$W_t = W_0 + \sum_{i=1}^M X_i(0,t] - \int_0^t \mathbb{1}_{[W_s>0]}ds$$

Using the Lebesgue-Stieltjes integration by parts formula we have:

$$W_t^2 = W_0^2 + 2\int_0^t W_{s-}dW_s + \sum_{i=1}^M \sum_{j=1}^{A_{i,t}} \sigma_{i,j}^2$$

By the assumptions we can assume the queue is in the stationary state and taking expectations we obtain:

$$\sum_{i=1}^M \mathbb{E}[\int_0^t W_{s-}\sigma_{i,s}dA_{i,s}] - \mathbb{E}[\int_0^t W_s \mathbb{1}_{[W_s>0]}ds] + \sum_{i=1}^M \mathbb{E}[\sum_{j=1}^{A_{i,t}} \sigma_{i,j}^2] = 0$$

By the definition of the Palm measures and the stationarity of W_t we obtain:

$$2\mathbb{E}[W_0] = 2\sum_{i=1}^M \lambda_i \mathbb{E}_{A_i}[W_{0-}\sigma_{i,0}] + \sum_{i=1}^M \lambda_i \mathbb{E}_{A_i}[\sigma_{i,0}^2] \tag{2.35}$$

If we specialize the formula to the $GI/GI/1$ case then we see $\mathbb{E}_{A_i}[W_{0-}\sigma_{i,0}] = \rho_i \mathbb{E}_{A_i}[W_{0-}]$ and hence we obtain that

$$\sum_{i=1}^M \rho_i (\mathbb{E}_{A_i}[W_{0-}] + \frac{1}{2}\mathbb{E}_{A_i}[R_i]) = \mathbb{E}[W_0]$$

where R_i is the stationary residual time of $\sigma_{i,0}$.

When specialized to the $M/G/1$ case we see $\mathbb{E}_{A_i}[W_{0-}] = \mathbb{E}[W_0]$ by PASTA and then we see that:

$$\mathbb{E}[W_0] = \frac{\sum_{i=1}^M \rho_i \mathbb{E}[R_i]}{2(1 - \sum_{i=1}^M \rho_i)}$$

which is just the P-K formula obtained for a $M/G/1$ queue with an input as a Poisson process with rate $\sum_{i=1}^M \lambda_i$ and the work brought has the distribution $\sigma = \sigma_i$ with probability $\frac{\lambda_i}{\sum_{i=1}^M \lambda_i}$. Moreover, it is also the mean workload seen by each type of arrival.

With this we conclude the discussion of the main formulae associated with mean queue lengths and waiting times.

2.4   FROM MEANS TO DISTRIBUTIONS

So far we have concentrated on the computation of means. However, the RCL also turns out to be a very effective tool to compute distributions of queues which we turn our attention to in this section.

2.4.1 EQUILIBRIUM DISTRIBUTIONS

Let us first begin by computing the stationary distribution of the forward recurrence time. This is called the equilibrium distribution of R_0. In Chapter 1, the expression for the mean of the forward recurrence time is to be computed under the Palm distribution. In the $M/G/1$ case because of PASTA we obtain the mean under the stationary distribution.

Let $R_t = T_+(t) - t$ where by definition $T_+(t) = T_{N_t+1}$ where $\{T_n\}$ are the points of a stationary point process N_t. Define: $X_t = (R_t - x)^+$.

Then $X_t^+ = -\mathbb{1}_{[R_t > x]}$ and the jumps of X_t are given by: $\Delta X_{T_n} = (R_{T_n} - x)^+ - (R_{T_{n-}} - x)^+$. By definition $R_{T_n} = T_{n+1} - T_n$ and $R_{T_{n-}} = 0$ and therefore for any $x \geq 0$ we have

$$\Delta X_{T_n} = (T_{n+1} - T_n - x)^+$$

Applying the RCL we obtain:

$$\mathbb{E}[\mathbb{1}_{[R_0 > x)]}] = \lambda_N \mathbb{E}_N[(T_1 - x)^+] = \lambda_N \int_x^\infty (y - x) dF(y)$$

where $F(x)$ is the distribution of the inter-arrival time $S_0 = T_1 - T_0 = T_1$ and under $\mathbb{P}_N$ we have $T_0 = 0$.

Let us define $\bar{F}_e(x) = \lambda_N \mathbb{E}_N[(T_1 - x)^+]$, then it is easy to see that $\bar{F}_e(0) = 1$, $\lim_{x \to \infty} \bar{F}_e(x) = 0$, and thus $F_e(x) = 1 - \bar{F}_e(x)$ defines the distribution of a r.v. that we call R_e. $F_e(x)$ is called the equilibrium distribution of R_0 with density $\lambda_N(1 - F(x))$ where $F(x)$ is the distribution of T_1 under $\mathbb{P}_N$. By direct calculation $\mathbb{E}[R_e] = \int_0^\infty \bar{F}_e(x) dx = \frac{1}{2} \mathbb{E}_N[T_1^2]$.

A similar argument can be carried out for the backward recurrence time denoted by B_0 and the stationary distribution is also equal to $F_e(x)$. Let us study the distribution of the equilibrium distribution a bit further. Let $S_e = R_e + B_e$ where R_e is the stationary forward recurrence time and B_e is the stationary backward recurrence time. S_e is called the equilibrium spread or lifetime of the point process.

By definition the spread is defined as $S_t = T_{N_t+1} - T_{N_t}$.

Therefore:

$$\frac{1}{t} \int_0^t \mathbb{1}_{[S_u > x]} du = \frac{1}{t} \sum_{k=1}^{N_t} S_k \mathbb{1}_{[S_k > x]}$$

i.e, we count all the inter-arrival times that are larger than x.

Hence, noting that $\lim_{t \to \infty} \frac{N_t}{t} = \lambda_N$ we obtain that $\mathbb{P}(S_e > x) = \lim_{t \to \infty} \frac{1}{t} \int_0^t \mathbb{1}_{[S_s > s]} ds$ satisfies:

$$\mathbb{P}(S_e > x) = \lambda_N \mathbb{E}_N[T_1 \mathbb{1}_{[T_1 > x]}]$$

from which it readily follows that:

$$\mathbb{E}[S_e] = \mathbb{E}[R_e] + \mathbb{E}[B_e] = \frac{\mathbb{E}_N[T_1^2]}{\mathbb{E}[T_1]}$$

Proposition 2.19 *Let $S_e = R_e + B_e$, and $F_S(x)$ denote the equilibrium distribution of a point process whose Palm distribution is given by $F(x) = \mathbb{P}_N(T_1 \leq x)$. Then S_e is stochastically larger (under $\mathbb{P}$) than $T_1 - T_0$ under $\mathbb{P}_N$.*

Proof. Let $F_S(x) = \mathbb{P}(S_e \leq x)$. To show this result we need to show that:

$$1 - F_S(x) = \mathbb{P}(S_e > x) \geq 1 - F(x) \quad x \geq 0$$

Now:

$$\mathbb{P}(S_e > x) = \lambda_N E_N[T_1 \mathbb{1}_{[T_1 > x]}] = \lambda_N x(1 - F(x)) + (1 - F_e(x))$$

Noting that: $\lambda_N = \frac{1}{\int_0^\infty (1 - F(y)) dy}$. We therefore need to show $x(1 - F(x)) + \frac{1}{\lambda_N}(1 - F_e(x)) \geq \frac{1}{\lambda_N}(1 - F(x))$.

This readily follows from

$$
\begin{aligned}
[\lambda_N]^{-1}(1 - F(x)) &= (1 - F(x)) \int_0^x (1 - F(y)) dy + (1 - F(x)) \int_x^\infty (1 - F(y)) dy \\
&\leq x(1 - F(x)) + \lambda_N^{-1}(1 - F_e(x))
\end{aligned}
$$

From which the result follows. $\square$

The interpretation of an equilibrium distribution is as follows: suppose we define a point process N_t^e whose points T_n^e are such that $S_n^e = T_{n+1}^e - T_n^e$ are i.i.d. and distributed according to $F_e(x)$ then N_t^e is $\mathbb{P} \circ \theta_t$ invariant. In the simplest case the Poisson process only has this property. The fact that S_e is stochastically greater than $S_0 = T_1 - T_0$ is a manifestation of the *inspection paradox*.

2.4.2 WORKLOAD AND BUSY PERIOD DISTRIBUTIONS

Let W_t be the stationary workload of a $G/G/1$ queue.

Define $X_t = e^{-hW_t}$. Then: $X_t^+ = he^{-hW_t}\mathbb{1}_{[W_t > 0]}$ and

$$\Delta X_{T_n} = e^{-h(W_{T_n-} + \sigma_n)} - e^{-hW_{T_n-}} = e^{-hW_{T_n-}}[e^{-h\sigma_n} - 1]$$

Applying the RCL we obtain:

$$\mathbb{E}[e^{-hW_0}\mathbb{1}_{[W_0 > 0]}] = \frac{\lambda_A}{h}\mathbb{E}_A[e^{-hW_0-}[e^{-h\sigma_0} - 1]]$$

When specialized to the $GI/GI/1$ case, we have σ_0 is independent of W_{0-}, and noting that $\mathbb{P}(W_0 > 0) = \rho = \lambda_A \mathbb{E}_A[\sigma_0]$ we obtain:

$$\mathbb{E}[e^{-hW_0}] = (1 - \rho) + \frac{\lambda_A}{h} \mathbb{E}_A[e^{-h\sigma_0} - 1]\mathbb{E}_N[e^{-hW_{0-}}] \qquad (2.36)$$

In the $M/G/1$ case because of PASTA we have $\mathbb{E}_A[e^{-hW_{0-}}] = \mathbb{E}[e^{-hW_0}]$ and we obtain:

$$\mathbb{E}[e^{-hW_0}] = \frac{h(1 - \rho)}{h - \lambda \mathbb{E}[e^{-h\sigma_0} - 1]} = \frac{h(1 - \rho)}{h - \lambda(F^*(h) - 1)} \qquad (2.37)$$

where $F^*(h) = \mathbb{E}[e^{-h\sigma_0}]$ is the moment generating function of σ_0. This last equation is referred to as Takacs' equation for $M/G/1$ queues.

A busy period of a queue is defined as the duration from the instant an arrival occurs to an empty queue up to the time when the workload hits 0 again. Knowing the length of a busy period of a queue is a very useful quantity because it allows us to determine how many packets have been processed, etc. In general, determining its distribution is not easy but in the $M/G/1$ case we can exploit the fact that the total work brought into the system during an interval of length t has stationary independent increments, a so-called Levy process. This property implies that the arriving work is a Markov process with independent increments. Once again, let us assume that the server serves at unit rate.

Definition 2.20 The busy period is defined as:

$$B_0 = \inf_t \{t > 0 : W_t = 0 | W_0 = \sigma_0\} = \inf_t \{t > 0 : \sigma_0 + \int_0^t \sigma_s dA_s - t \leq 0\} \qquad (2.38)$$

where $\int_0^t \sigma_s dA_s = \sum_{n=1}^{A_t} \sigma_n$.

Under the assumption that $\rho < 1$ the sequence $\{B_n\}$, where B_i denotes the ith busy period, is i.i.d. by the stationary independent increment property of the input. Let us examine the behavior of a busy period: an arrival brings σ_0 and during the time it takes to process this work $A(0, \sigma_0]$ arrivals take place and each arrival generates a new sub-busy period, etc. Thus:

$$B \stackrel{d}{=} \sigma_0 + \sum_{i=1}^{A(0,\sigma_0]} B_i$$

where B_i are the busy periods generated by each of the arrivals $A(0, \sigma_0]$

From the definition and the use of Wald's equation we readily obtain that:

$$\mathbb{E}[B_0] = \frac{\mathbb{E}[\sigma_0]}{1 - \rho} \qquad (2.39)$$

We can also explicitly calculate the Laplace transform of the busy period as follows:

$$
\begin{aligned}
B^*(h) &= \mathbb{E}[e^{-hB}] \\
&= \mathbb{E}[\mathbb{E}[e^{-h(x+\sum_{i=1}^{A(0,x]} B_i)}|\sigma_0 = x]] \\
&= \int_0^\infty \mathbb{E}[e^{-hx}[B^*(h)]^{A(0,x]}]dF(x) \\
&= \int_0^\infty e^{-hx-\lambda x(1-B^*(h))}dF(x) \\
&= F^*(h+\lambda(1-B^*(h))
\end{aligned}
$$

where $B^*(h) = \mathbb{E}[e^{-hB}]$ and $F^*(h) = \mathbb{E}[e^{-h\sigma_0}]$ denote the Laplace transforms of B and σ_0 respectively.

This is an implicit equation for solving for $B^*(h)$. We now show that if $\rho < 1$ then we can obtain a unique solution in terms of the Laplace transform of σ_0.

Define: $f(h) = h + \lambda(F^*(h) - 1)$. Then it is easy to see that $f(0) = 0$ and $f(1) = 1 + \lambda[\mathbb{E}[e^{-\sigma} - 1]$. Moreover, $f''(h) > 0$ implying it is strictly convex, $F'(h)|_{h=0} = 1 - \rho \in (0, 1)$, and hence $\exists\, h^*$ which is unique such that $f(h^*) = h^*$ i.e., there exists a unique fixed point in $(0, \infty)$. Let us denote by $h^* = f^{-1}(h)$. Then :

$$
B^*(h) = F^*(f^{-1}(h)) \tag{2.40}
$$

To see this, from the definition of $f^{-1}(h)$ we have $B^*(f(h)) = F^*(h)$. Now $f^{-1}(h) = h + \lambda(1 - F^*(f^{-1}(h))) = h + \lambda(1 - B^*(h))$ which is equivalent to $B^*(h) = F^*(h + \lambda(1 - B^*(h))$ as required.

2.4.3   STATIONARY DISTRIBUTIONS OF $GI/M/1$ QUEUES

We conclude our discussion with a discussion of $GI/M/1$ queues where the arrival process is a renewal process whose inter-arrival times are i.i.d. with $\mathbb{E}_N[T_{n+1} - T_n] = \frac{1}{\lambda_A}$ and the amount of work brought is i.i.d. exponential with $\mathbb{E}[\sigma] = \frac{1}{\mu}$. We assume $\lambda_A \mathbb{E}[\sigma] = \rho < 1$. The standard technique for analyzing such queues is to consider the queueing process at arrival times. Let $Q_n = Q_{T_n-}$ Then the evolution of the queue is as follows:

$$
Q_{n+1} = Q_n + 1 - D(T_n, T_{n+1}]
$$

where $D(T_n, T_{n+1}]$ denotes the number of departures in $(T_n, T_{n+1}]$ Now $D(T_n, T_{n+1}) \le Q_n + 1$ since the number of departures between 2 arrivals cannot exceed the total number in the queue. Let $H(t) = \mathbb{P}_A(T_1 \le t)$ denote the inter-arrival time distribution and under $\mathbb{P}_A$ the inter-arrival times are i.i.d. The departure distribution can be computed explicitly from:

$$
p_k = \mathbb{P}(D(T_n, T_{n+1}] = k) = \int_0^\infty \frac{(\mu t)^k}{k!}e^{-\mu t}dH(t)
$$

since the event corresponds to k-departures in one inter-arrival time and the departures are Poisson since the service times are exponential.

Since $D(T_n, T_{n+1}]$ are i.i.d. it follows that $\{Q_n\}$ is a Markov chain on $\{0, 1, 2, \cdots\}$ with the following transition structure:

$$P_{i,j} = p_k \quad j = i+1-k, \quad k = 0, 1, 2, \ldots, i$$

$$P_{i,0} = 1 - \sum_{k=0}^{i} p_k$$

Define the moment generating function $P(z)$ of $\{p_k\}$:

$$P(z) = \sum_{k=0}^{\infty} p_k z^k \tag{2.41}$$

$$= \sum_{k=0}^{\infty} z^k \int_0^{\infty} \frac{(\mu t)^k}{k!} e^{-\mu t} dH(t)$$

$$= \phi_A(\mu(1-z)), \tag{2.42}$$

where $\phi_A(z) = \mathbb{E}_A[e^{-z T_1}]$ (noting that $\mathbb{P}_A(T_0 = 0) = 1$ by definition of the Palm probability). Then we can state the following result.

Proposition 2.21   *Consider a $GI/M/1$ queue in equilibrium with $\lambda_A \mathbb{E}_A[\sigma_0] = \rho < 1$. Then there exists a unique solution in $(0, 1)$ to the fixed point equation:*

$$\xi = P(\xi) \tag{2.43}$$

where $P(z)$ is defined as in (2.41). Moreover, the arrival stationary distribution of the queue is given by:

$$\boxed{\pi_A(i) = \xi^i (1 - \xi), \quad i = 0, 1, 2, \ldots} \tag{2.44}$$

Proof. Let us show that $\pi_A(i)$ as defined is the arrival stationary distribution, i.e., we need to show $\pi_A = \pi_A P$ where P is the transition matrix of the chain that results at arrival instants. Now:

$$\pi_A(j) = \sum_{i=0}^{\infty} \pi_A(i) P_{i,j}$$

$$\pi_A(0) = \sum_{i=0}^{\infty} \pi_A(i) P_{i,0}$$

By definition $\pi_A(0) = (1 - \xi)$ and therefore:

$$1 - \xi = \sum_{i=0}^{\infty} (1 - \xi)\xi^i (1 - \sum_{j=0}^{i} p_j)$$

which is equivalent to:

$$1 - \frac{1}{1 - \xi} = -\sum_{i=0}^{\infty} \xi^i \sum_{j=0}^{i} p_j$$

$$\frac{\xi}{1 - \xi} = \sum_{j=0}^{\infty} p_j \sum_{i=j}^{\infty} \xi^i$$

$$= \frac{P(\xi)}{1 - \xi}$$

Hence, we have shown $\pi_A(i) = \xi^i(1 - \xi) \Leftrightarrow \xi = P(\xi)$. It remains to show that if $\rho < 1$ there exists a unique solution in $(0, 1)$.

To see this first note that $P(0) = p_0 = \phi_A(\mu) > 0$ and $P(1) = \phi_A(0) = 1$. Furthermore, $P''(z) = \mu^2 \phi_A(\mu(1 - z)) > 0$ for $z \in (0, 1)$ implying $P(z)$ is convex in $(0, 1)$. Finally, noting that $P'(z)_{z=1} = \mu \mathbb{E}_A[T_1] = \rho^{-1} > 1$ if $\rho < 1$. This implies that there exists a $\xi \in (0, 1)$ such that $P(\xi) = \xi$. Uniqueness follows from strict convexity.

$\square$

From the form of the arrival stationary distribution and the fact that the services are exponentially distributed, the mean sojourn time of packets in the system is:

$$\mathbb{E}_A[W_0] = \sum_{i=1}^{\infty} (i + 1)\pi_A(i)\mathbb{E}[\sigma_0]$$

$$= \frac{\mathbb{E}[\sigma_0]}{1 - \xi} \tag{2.45}$$

Noting that the average arrival rate to the queue is $\lambda_A = (\mathbb{E}_A[T_1])^{-1}$ we obtain from Little's law:

$$\mathbb{E}[Q_0] = \frac{\lambda_A \mathbb{E}[\sigma_0]}{1 - \xi} = \frac{\rho}{1 - \xi} \neq \frac{\xi}{1 - \xi} = \mathbb{E}_A[Q_0]$$

Having obtained the mean waiting time (under FIFO) we can show the following property of $GI/M/1$ queues.

Proposition 2.22 *Amongst all the input distributions for the inter-arrival times in $GI/M/1$ queues in equilibrium that result in the same average load ρ, the deterministic distribution corresponding to arrival times spaced $\mathbb{E}_A[T_1] = \frac{1}{\lambda_A}$ minimizes the mean sojourn times in the queue.*

Proof. By the convexity of the exponential function:

$$\phi_A(h) = \mathbb{E}_A[e^{-hT_1}] \geq e^{-h\mathbb{E}_A[T_1]} = e^{-\frac{h}{\lambda_A}} = \phi_D(h)$$

where $\phi_D(h)$ is the Laplace transform of the deterministic inter-arrival time of length λ_A^{-1}. Hence, for any $z \in (0, 1)$ we have $P_{G/M/1}(z) = \phi_A(\mu(1-z)) \geq P_{D/M/1}(z) = \phi_D(\mu(1-z))$.

Let $\xi_1 = P_{G/M/1}(\xi_1)$ and $\xi_2 = P_{D/M/1}(\xi_2)$ and by the property above it implies $\xi_1 \geq \xi_2$ and from the mean of the sojourn time it follows that $\mathbb{E}_{G/M/1}[W_0] \geq \mathbb{E}_{D/M/1}[W_0]$ □

Remark 2.23

Earlier in the context of the Pollaczek-Khinchine formula for $M/G/1$ queues we had shown that $\mathbb{E}_{M/G/1}[W_0] \geq \mathbb{E}_{M/D/1}[W_0]$. Thus, combining these insights we see that average waiting times in queues are reduced by making the arrivals less variable (deterministic) and making packet lengths less variable (deterministic). This is referred to as *determinism minimizes waiting time in queues.*

We conclude our discussion by showing the relation between the arrival stationary and (time) stationary distribution. This follows from equations (2.19) and (2.22) noting the fact that the service times are exponentially distributed. Indeed, from the fact that the arrivals are i.i.d. we have:

$$\lambda_A \pi_A(i) = \mu \pi_D(i+1) = \mu \pi(i+1), \quad i = 0, 1, \ldots$$

since $\pi_D(i+1) = \pi(i+1)$ from the fact that the rate associated with exponential distributions is constant (PASTA). Therefore, noting that $\rho = \frac{\lambda_A}{\mu}$ we have:

$$\pi(i+1) = \rho \pi_A(i), \quad i = 0, 1, 2 \ldots$$

and from the normalization condition $\sum_{i=0}^{\infty} \pi(i) = 1$ we finally obtain the stationary distribution as:

$$\pi(i) = \rho(1-\xi)\xi^{i-1}, \quad i = 1, 2, \ldots \tag{2.46}$$
$$\pi(0) = 1 - \rho \tag{2.47}$$

With this computation we can now verify Little's formula $\mathbb{E}[Q_0] = \lambda_A \mathbb{E}_A[W_0]$ holds.

Finally, from this computation we can identify the conditional intensities in equation (2.22) as:

$$\lambda_i = \frac{\lambda_A \xi}{\rho}, \quad i = 1, 2, \ldots$$
$$\lambda_0 = \lambda_A \frac{1-\xi}{1-\rho}$$

Finally, we note that if the arrival process is Poisson then $H(t) = 1 - e^{-\lambda t}$, and therefore $\xi = \rho$ and we recover the result that $\pi(i) = \pi_A(i)$ as is well known for $M/M/1$ queues.

2.4.4   OUTPUT PROCESSES OF QUEUES

In our discussion so far we have analyzed single queues in isolation. However, in applications the most common model is that of an interconnection of queues to form a queueing network. In order to be able to analyze this situation it is necessary to understand the stochastic process that characterizes the departure process of queues. However, in this context there are very few results for queues of the $G/G/.$ type. However, there are useful results that allow us to analyze queueing networks in two cases, namely the $M/M/s$ and the $M/G/\infty$ situations. Below we present the results for the $M/M/1$ case. The $M/G/\infty$ case will be discussed in the next chapter.

Consider an $M/M/1$ queue in equilibrium with arrival rate λ and service rate μ with $\rho = \frac{\lambda}{\mu} < 1$. Let D_n denote the departure instants of packets in the queue.

Let $D^*(h) = \mathbb{E}[e^{-h(D_{n+1}-D_n)}]$ denote the Laplace transform of the inter-departure times. Then:

$$D^*(u) = \mathbb{E}[e^{-h(D_{n+1}-D_n)}|Q_{D_n} > 0]\mathbb{P}(Q_{D_n} > 0) + \mathbb{E}[e^{-h(D_{n+1}-D_n)}|Q_{D_n} = 0]\mathbb{P}(Q_{D_n} = 0)$$

Noting that $\mathbb{P}(Q_{D_n} = i) = \pi_D(i) = \pi_A(i) = \pi(i)$ by PASTA we have $\mathbb{P}(Q_{D_n} > 0) = \rho$ and if the system is non-empty then the inter-departure times are $\sim \exp(\mu)$ and so

$$\mathbb{E}[e^{-h(D_{n+1}-D_n)}|Q_{D_n} > 0] = \frac{\mu}{h+\mu}$$

Now the second term:

$$\begin{aligned}\mathbb{E}[e^{-h(D_{n+1}-D_n)}|Q_{D_n} = 0] &= \mathbb{E}[e^{-h(D_{n+1}-T_{n+1}-D_n+T_{n+1})}|Q_{D_n}=0]\\ &= \mathbb{E}[e^{-h(X_n+Y_n)}|Q_{D_n} = 0]\end{aligned}$$

where $X_n \sim D_{n+1} - T_{n+1}$ and $Y_n \sim T_{n+1} - D_n$ and by the memoryless property of the exponential distributions (conditioned on independent random times) we have $X_n \sim \exp(\mu)$ and $T_{n+1} - D_n \sim \exp(\lambda)$ from the fact that on $Q_{D_n} = 0$ we have $D_{n+1} > T_{n+1} > D_n$. Moreover, by the independent increment property X_n and Y_n are independent.

Therefore, we obtain:

$$\begin{aligned}D^*(h) &= \rho\frac{\mu}{h+\mu} + (1-\rho)\frac{\mu}{h+\mu}\frac{\lambda}{h+\lambda}\\ &= \frac{\lambda}{h+\lambda}\end{aligned}$$

This is just the Laplace transform of an exponentially distributed r.v. with rate λ and hence $D(0,t] = \sum_n \mathbb{1}_{[0<D_n\leq t]}$ is a Poisson process with rate λ same as the input process. From the independent increment property of the Poisson process $D(s,t]$ is independent of $Q_u, \ u \leq s$. Actually one can even say more $\mathbb{P}(Q_t = n|D(0,t]) = \pi(n)$ for $n \geq 0, t \geq 0$. This important result is called Burke's theorem which is stated in more generality below:

Proposition 2.24   *The departure process of an $M/M/c$ queue in equilibrium is a Poisson process with the same intensity λ as the arrival process. Moreover, $\mathbb{P}(Q_\tau = n|\mathcal{F}_\tau^D) = \pi(n)$ where τ is any $\mathcal{F}_t^D$ stopping time where $\mathcal{F}_t^D$ is the filtration or history of the departures up to t.*

2.5 FLUID QUEUES

Fluid queues or rather queues with continuous arrival processes are natural models in the context of today's high-speed networks where the line rates are so high that the granularity of packets is lost and the input can be seen as a flow of bits at a given rate. Congestion control mechanisms such as TCP being the mechanism for transfer of bits results in inputs with variable rates. The continuous nature of the inputs results in queues whose state-space is just a non-negative real value and thus from a modeling standpoint the workload model is most appropriate where the input now will denote the total amount of information (bits) arriving in a given interval. In this section we develop the queueing analysis for fluid queues, where the goal is to obtain the analog of Little's formula and the Pollaczek-Khinchine formula. As mentioned in Chapter 1 we now have to work with the stationary distributions and the fluid analog of the Palm measure.

Consider a work conserving queue with a server that works at a constant rate c when there is work in the queue. Let $A(0, t]$ denote the total amount of work (or bits) brought into the queue in the interval $(0, t]$ that is assumed to be a continuous process with stationary increments and let $\lambda_A = \mathbb{E}[A(0, 1]$. Let $\{W_t\}$ denote the content of the queue at time t or the workload. Then as before:

$$W_t = W_0 + A(0, t] - ct + I_t \tag{2.48}$$

where $I_t = c \int_0^t \mathbb{1}_{[W_s=0]} ds$ denotes the idle time process. Noting that $\int_0^t W_s dI_s = 0$ we have for function $f \in C^1$:

$$f(W_t) = f(W_0) + \int_0^t f'(W_s) \mathbb{1}_{[W_s>0]} d(A_s - cs) \tag{2.49}$$

Then from the definition of the stationary Palm measure in Chapter 1 we have for any stationary process $X_s = X_0 \circ \theta_s$:

$$\mathbb{E}[\int_0^t X_s dA_s] = \lambda_A t \mathbb{E}_A[X_0] \tag{2.50}$$

Now by the characterization of the regulator process I_t and using similar arguments as in the proof of the stability of queues with stationary point process inputs we can show that under the assumption that $\rho_A = \lambda_A c^{-1} < 1$ there exists a stationary version of $\{W_t\}$ and is given by:

$$W_0 = \sup_{t \geq 0} \{A(-t, 0] - ct\}^+ \tag{2.51}$$

So in the following we will assume that $\rho_A < 1$ and the queue is stationary (or in equilibrium). Then we can show the following results.

Proposition 2.25 *Assume $\rho_A = \lambda_A c^{-1} < 1$ and that the queueing process is in equilibrium. Then:*

1. For all continuous functions φ

$$\mathbb{E}[\varphi(W_0) \mathbb{1}_{[W_0>0]}] = \rho_A \mathbb{E}_A[\varphi(W_0) \mathbb{1}_{[W_0>0]}] \tag{2.52}$$

2. For all Borel sets $B \in \mathfrak{R}$ which do not contain the origin:

$$\mathbb{P}(W_0 \in B) = \rho_A \mathbb{P}_A(W_0 \in B) \tag{2.53}$$

and at the origin 0, i.e., $B = \{0\}$:

$$\mathbb{P}(W_0 = 0) = \rho_A[\mathbb{P}_A(W_0 = 0) - 1] + 1 \tag{2.54}$$

3. A Little's law for fluid queues is given by:

$$\mathbb{E}[W_0] = \rho_A \mathbb{E}_A[W_0] \tag{2.55}$$

Proof. The proof follows directly from equation (2.49), stationarity, and the definition of the fluid Palm measure given by (2.50).

For the first result choose $f'(.) = \varphi(.)$. Then we have: $\mathbb{E}[f(W_t)] = \mathbb{E}[f(W_0)]$ and then noting $f'(.) = \varphi(.)$ we have from (2.49):

$$
\begin{aligned}
c\mathbb{E}[\int_0^t \varphi(W_s)ds] = ct\mathbb{E}[\varphi(W_0)\mathbb{1}_{[W_0>0]}] &= \mathbb{E}[\int_0^t \varphi(W_s)\mathbb{1}_{[W_s>0]}dA_s] \\
&= t\lambda_A \mathbb{E}_A[\varphi(W_0)\mathbb{1}_{[W_0>0]}]
\end{aligned}
$$

To show the second result we choose $f'(.) = \mathbb{1}_{[W_0 \in B]}$ and the equation for the origin follows from the fact that $\mathbb{1}_{[W_0=0]} = 1 - \mathbb{1}_{[W_0>0]}$.

Finally, Little's formula follows by choosing $\varphi(x) = x$. $\square$

Remark 2.26   In this fluid case, Little's formula serves directly as an inversion formula since the quantities on the left-hand side and the right-hand side are the same except that they are calculated under different measures.

We now obtain the fluid equivalent of the Pollaczek-Khinchine equation for the mean workload under the assumption that the input is of the ON-OFF type. Let us first notice one aspect of the fluid queueing model, if the rate $\frac{dA_t}{dt} < c$ for all t then the workload will always be 0 (assuming the queue is initially empty) or will become 0 after some finite time and remain 0. Thus, we need the rate $\frac{dA_t}{dt} > c$ for a fraction of time which is non-zero to have queueing occur and the probability that the queue is non-zero is non-zero. This is referred to as the *burstiness* assumption. We will formalize this in the assumptions below.

Consider the input $A(0, t]$ as shown in Figure 2.2 to be as defined by (1.23) with the following assumptions:

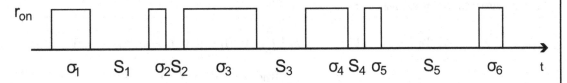

Figure 2.2: Fluid ON-OFF source with constant rates.

(A1) Let $T_n = \sigma_n + S_n$ denote the length of an ON+OFF period with $\{\sigma_n\}$ being an i.i.d sequence and $\{S_n\}$ being i.i.d. exponentially distributed r.v's with mean $(\lambda)^{-1}$. Let N denote the point process associated with the points $\{T_n\}$.

(A2) The source transmits at a rate r_t such that $r_t = r > c$ for $t \in (T_n, T_n + \sigma_n]$ and $r_t = 0$ for $t \in [T_n + \sigma_n, T_{n+1})$, i.e., it transmits at a fixed rate r when ON and is silent in the OFF periods $\{S_n\}$.

(A3): $\lambda_A p = \mathbb{E}_N[A(0, T_1]] < 1$ and $q = 1 - p = \mathbb{P}(T_0 + \sigma_0 < 0) = \mathbb{P}(0 \in S_0)$

Assumption (A2) corresponds to the burstiness assumption which states that during an ON period the source brings in more work than that which can be served in that interval. Define $\xi_t = \sum_{n=-\infty}^{\infty} \mathbb{1}_{[t \in [T_n, T_n + \sigma_n]]}$ i.e., $\xi_t = 1$ if t falls during an ON period of the source.

We now obtain the mean buffer content under the stationary distribution that corresponds to the Pollaczek-Khinchine formula for fluid queues.

Proposition 2.27 *Consider a fluid queue with a server of rate c with an ON-OFF arrival process which arrives at a rate $r > c$ when the arrival process is in the ON state and has a rate 0 in the OFF state. Let $\{\sigma_i\}$ denote the ON periods which are assumed to be i.i.d. with $\mathbb{E}[\sigma_0^2] < \infty$ and $\{S_0\}$ the silent periods are exponentially distributed with rate λ such that: $r \frac{\mathbb{E}[\sigma_0]}{\mathbb{E}[\sigma_0 + S_0]} = \mathbb{E}[A(0, 1]] = \lambda_A < c$, then the mean buffer content under the stationary distribution is given by:*

$$\mathbb{E}[W_0] = \frac{1}{c - \lambda_A}\left[\lambda_A \frac{\mathbb{E}_N[\sigma_0^2]}{2\mathbb{E}[\sigma_0]}(r - c)(1 - p)\right] \quad (2.56)$$

where $\lambda_A = \mathbb{E}[A(0, 1]]$ and N_t denotes the point process which counts the beginning of ON periods of the arrival process, and $p = \frac{\mathbb{E}_N[\sigma_0]}{\mathbb{E}_N[\sigma_0] + \mathbb{E}_N[S_0]}$.

Proof. By definition $p = \mathbb{E}[\xi_0]$ and with the convention $T_0 < 0 < T_1 < \ldots$, we obtain :

$$\mathbb{E}[W_0] = \mathbb{E}[W_0 \mathbb{1}_{[\xi_0 = 1]}] + \mathbb{E}[W_0 \mathbb{1}_{[\xi_0 = 0]}] \quad (2.57)$$

On the event that $\xi_0 = 1$ it means the queue is never empty on $[T_0, 0]$, and thus:

$$W_0 = W_{T_0} + A(T_0, 0] + cT_0$$

The event $\xi_0 = 1$ is equivalent to $\mathbb{1}_{[T_0 + \sigma_0 \geq 0]}$. Therefore:

$$
\begin{aligned}
\mathbb{E}[W_0 \mathbb{1}_{[T_0 + \sigma_0 \geq 0]}] &= \lambda_N \mathbb{E}_N[\int_0^{T_1} (W(0) + A(0,t] - ct)\mathbb{1}_{[\sigma_0 \geq t]})dt] \\
&= \lambda_N \mathbb{E}_N[W_0 \sigma_0] + \lambda_N \mathbb{E}_N[\int_0^{\sigma_0} (r - c)t \, dt] \\
&= \lambda_N \mathbb{E}_N[\sigma_0]\mathbb{E}_N[W_0] + \lambda_N (r - c)\frac{\mathbb{E}_N[\sigma_0^2]}{2} \\
&= p\mathbb{E}_N[W_0] + \lambda_N (r - c)\frac{\mathbb{E}_N[\sigma_0^2]}{2}
\end{aligned}
$$

by the Palm inversion formula and the fact that under $\mathbb{P}_N$ we have $T_0 = 0$ and σ_0 and W_0 are independent under $\mathbb{P}_N$. By definition $\lambda_N \mathbb{E}[\sigma_0] = p = \frac{\lambda_A}{r}$

We need to relate $\mathbb{E}_N[W_0]$ to the expectation under the Palm measure of the arrival process. For this we use Neveu's exchange formula (even though in this case $A.$ is not a point process)

$$
\lambda_A \mathbb{E}_A[W_0] = \lambda_N \mathbb{E}_N[\int_0^{T_1} W_s \, dA_s]
$$

Noting that $dA_s = r\mathbb{1}_{[T_0 < s \leq T_0 + \sigma_0]}ds$ and using the fluid Little's formula $\lambda_A \mathbb{E}_A[W_0] = c\mathbb{E}[W_0]$ we obtain:

$$
\begin{aligned}
\lambda_A \mathbb{E}_A[W_0] &= \mathbb{E}_N[\int_0^{\sigma_0} (W_0 + rs - cs)r \, ds] \\
&= \lambda_A \mathbb{E}_N[W_0] + r(r - c)\lambda_N \frac{\mathbb{E}_N[\sigma_0^2]}{2}
\end{aligned}
\tag{2.58}
$$

Therefore,

$$
\mathbb{E}[W_0 \mathbb{1}_{[T_0 + \sigma_0 \geq 0]}] = p\mathbb{E}_A[W_0]
$$

Now using the Palm inversion formula for the second term of (2.57) we obtain:

$$
\begin{aligned}
\mathbb{E}[W_0 \mathbb{1}_{[\xi_0 = 0]}] &= \lambda_N \mathbb{E}_N[\int_0^{T_1} W_s \mathbb{1}_{[\sigma_0 \leq s]}ds] \\
&= \lambda_N \mathbb{E}_N[\int_{\sigma_0}^{\sigma_0 + S_0} W_s \, ds] \\
&= \lambda_N \mathbb{E}_N[S_0]\mathbb{E}_N[W_0] \\
&= (1 - p)\mathbb{E}_N[W_0]
\end{aligned}
$$

Where in the last step we use the property that under $\mathbb{P}_N$ the r.v. S_0 is exponential and the last step is an application of Wald's identity .

Substituting for $\mathbb{E}_N[W_0]$ from equation (2.58) we obtain:

$$
\mathbb{E}[W_0] = \frac{1}{c - \lambda_A}\left[r(r - c)\lambda_N \frac{\mathbb{E}_N[\sigma_0^2]}{2}[1 - p]\right]
$$

as announced. $\square$

Remark 2.28

1. The formula above corresponds to the fluid version of the Pollaczek-Khinchine formula . This can readily be extended when M independent ON-OFF sources are multiplexed. Let $\{A_i(0,t)\}_{i=1}^M$ denote M independent ON-OFF sources with i.i.d. ON periods during which source i transmits at $r_i > c$ and the OFF periods are i.i.d. exponential. Let $\lambda_A = \sum_{i=1}^M \lambda_i$ where $\lambda_i = \mathbb{E}[A_i(0,1]]$. Let $\{N_i\}$ be the point processes associated with he beginning of the ON periods. Then:

$$\mathbb{E}[W_0] = \frac{1}{c - \lambda_A}\left(\sum_{i=1}^M \lambda_{A_i} \frac{\mathbb{E}_{N_i}[\sigma_{0,i}^2]}{2\mathbb{E}[\sigma_{0,i}]}(r_i - c + \lambda_A - \lambda_i)[1 - p_i]\right) \qquad (2.59)$$

where $p_i = \frac{\mathbb{E}_{N_i}[\sigma_{i,0}]}{\mathbb{E}_{N_i}[\sigma_{i,0}] + \mathbb{E}_{N_i}[S_{i,0}]} = \lambda_{A_i} r_i^{-1}$.

2. We can handle more complex types of ON behavior while requiring the *burstiness* condition to hold. In particular, we can consider the arriving number of bits from source A_i to be given by $F_i(t - Ti, n)\mathbb{1}_{[T_{i,n} < t \le T_{i,n} + \sigma_{i,n}]}$ with $F_i(t) > ct$. In his case the mean workload is given by:

$$\mathbb{E}[W_0] = \frac{\sum_{i=1}^M (C_i - \lambda_i D_i)}{c - \lambda_A} \qquad (2.60)$$

where:

$$C_i = \lambda_{N_i} \mathbb{E}_{N_i}\left[\int_0^{\sigma_{i,0}} (A(t) - ct)dA_i(t)\right] \qquad (2.61)$$

and

$$D_i = \lambda_{N_i} \mathbb{E}_{N_i}\left[\int_0^{\sigma_{i,0}} (A(t) - ct)dt\right] \qquad (2.62)$$

This concludes our discussion of fluid queues. We will use these results in Chapter 4 when we discuss statistical multiplexing.

CONCLUDING REMARKS

The results we have seen in this chapter form the bread-and-butter of the elementary results of queueing theory. The derivations given here have been non-standard, or rather modern, with the aim of showing that these results are just simple consequences of stationarity and the sample-path evolution of the various processes. Essentially, most results are seen as consequences of the RCL and the definition of the Palm probabilities. Every attempt has been made to show the measure under which the various computations need to be done. This is not emphasized in more elementary books but as we have seen we can obtain incorrect results if we do not use the right measure.

One aspect we have not considered at all are queueing networks. These will be discussed in Chapter 4. Queueing network models are rich but explicit results are known only for some special cases. One of the difficulties is in characterizing the departure process from queues that are not of the $M/M/.$ type. However, some models are indeed tractable such as so-called Jackson networks and more generally, so-called quasi-reversible networks, where the input processes are Poisson and the service times at nodes are also independent and exponentially distributed. For these models, the vector of congestion processes at each queue evolves as a Markov process and in the FIFO and Processor Sharing case can be shown to have a so-called *product-form* stationary distribution. However, the applicability of these models to many networking applications is questionable because the service times of a given packet at different queues are assumed independent even if the packet length does not change as it traverses the network in reality. One particularly useful and tractable network model is in the context of loss networks which are in the next chapter. Except in the Markovian case very little is known even about the stability of general networks and this continues to be an active research area.

NOTES AND PROBING FURTHER

As mentioned above, the principal queueing formulae, namely Little's law and the Pollaczek-Khinchine formulae along with the stationary distributions for $M/M/.$, $M/G//1$ and the $GI/G/1$ queues forms the basis of queueing theory and thus can be found many textbooks. However, they usually rely in Markov chain techniques (except for Little's law which is shown by a simple sample path argument) rather than the approach taken here. The principal reason in introducing the RCL was to show that the queueing formulae are really simple consequences of stationarity and the sample-path evolution equations. Thus, many of the results can be derived under very general hypotheses although when performing explicit calculations we need to impose some simplifying assumptions. This approach has been of increasing use by many researchers but the results have not widely percolated into standard course material. One of the goals was to introduce the reader to this world of looking at queueing theory from a sample-path viewpoint without bringing in too much technical machinery. Even so, the material may appear daunting to beginners but once the concepts have been understood, operationally the theory is fairly straightforward. One of the key ideas that has not been introduced has been the notion of stochastic intensity and the related martingale structures of point processes. This is a very powerful framework and a key result known as Papangelou's theorem allows us to tie in the martingale approach with the Palm approach. This often considerably simplifies the analysis but on balance it was felt that this would take us too far out from the objectives of this monograph. As mentioned earlier, besides the excellent monograph of Baccelli and Brémaud there is no systematic account of the framework that has been used in book form. In the reference list some pointers are provided to some original papers where many other issues related to queueing and the mathematical framework has been used in this chapter. Once again, the list is indicative and not comprehensive. The reader is encouraged to probe further in the references cited in the books and articles.

BOOK REFERENCES

Elementary but mathematical treatments of queueing theory and networks can be found in many books. However, the following are particularly nice for their treatment of some topics.

L. Kleinrock, *Queueing Systems, Vol. 1: Theory*, John Wiley, N.Y., 1975.

L. Kleinrock, *Queueing Systems: Vol. 2 Computer Applications*, John Wiley, N.Y., 1976

> Volume 1 is a classical reference to queueing models and presents the basic theory in the Markovian context. It remains a standard reference and is very comprehensive. Volume 2 is devoted to applications and there is a good discussion of the conservation laws for $M/G/1$ queues.

R.W. Wolff; *Stochastic Modelling and the Theory of Queues*, Prentice-Hall, N.J., 1989

> This has now become a very standard reference for courses in queueing theory. It is quite comprehensive and with an emphasis on renewal theory.

F. P. Kelly; *Reversibility and Stochastic Networks*, Wiley, London, 1979.

> This is a classical and comprehensive reference on Markovian queueing networks and the role of reversibility of the underlying Markov processes in determining their stationary distributions.

F. Baccelli and P. Brémaud:*Elements of Queueing Theory: Palm-Martingale Calculus and Stochastic Recurrences*, 2nd Ed., Applications of Mathematics 26, Springer-Verlag, N. Y., 2005

> As mentioned earlier this has now become a standard reference on Palm theory and queues. Most of the mathematical framework and results that have been presented can be found in this book. It is initially a daunting book but is extremely accessible once the initial mathematical preliminaries have been understood. A highly recommended book for probing further.

M. El-Taha and S. Stidham Jr., *Sample-path Analysis of Queueing Systems*, Kluwer Academic Publ., 1999

> This book presents an operational point of view of results in queueing and Palm probabilities in an ergodic framework mush as has been followed in these notes, especially in Chapter 1. It is useful to understand the theory formally.

P. Robert; Stochastic Networks and Queues, Springer, Applications in Mathematics 52, Springer-Verlag, Berlin , 2003.

> This book is also excellent for learning the mathematical modeling and analysis of networks. The book presents a very comprehensive treatment of network processes and is particularly useful for results on point processes and their functionals. A very nice treatment of the Skorohod reflection problem is also given. It is a highly recommended book for the mathematically inclined.

H.C. Chen and D.D. Yao, *Fundamentals of Queueing Networks: Performance, Asymptotics, and Optimization*, Applications of Mathematics 46, Springer-Verlag, NY, 2001.

This is a very comprehensive view of queueing networks from the fluid viewpoint. The primary focus of this book is on stability theory, heavy traffic limit theorems, and on scheduling. Of particular note is the treatment of the reflection map associated with the multi-dimensional Skorohod reflection problem.

P. Brémaud; *Point Processes and Queues: Martingale Dynamics*, Springer Series in Statistics, Springer-Verlag, N.Y., 1980

This is still the best reference on the martingale approach to queues and point processes. This book presents a very nice overview of flows in Markovian networks, etc.

M. Bramson, *Stability of Queueing Networks*, Lecture Notes in Mathematics, Vol 1950, Springer, Berlin, 2008

This is an extremely comprehensive treatment on the use of fluid models to study the stability of queueing networks. Although we have not discussed them in this Chapter, this book collects all the know-how on the subject of stability and moreover shows that in networks, FIFO and the "usual condition" of $\rho < c$ might not be sufficient for stability. Given the interest in these models this is an important reference on the topic.

JOURNAL ARTICLES

In addition to the journal articles listed in Chapter 1 the following are suggested to further understand the scope and framework of the mathematical approaches that have been used in this chapter.

R. M. Loynes, *The stability of a queue with non-independent inter-arrival and service times*, Proc. Cambridge Philos. Soc. 58 1962 497–520.

This paper is a classic and first presents the Loynes schema and how the stability of $G/G/1$ queues can be studied via the monotonicity properties of the map. The $\rho < 1$ condition being necessary and sufficient for stability is proved.

S. Foss, and T. Konstantopoulos, *An overview of some stochastic stability methods*, J. Oper. Res. Soc. Japan 47 (2004), no. 4, 275–303.

This is a very nice and comprehensive survey paper on the various tools to study stability of queues, namely the Loynes type of argument, arguments based on renovating events and coupling, and Lyapunov techniques based for Markov processes. In these notes we have not discussed coupling or renovation which are very powerful tools to study stability of recursive stochastic sequences that occur in Lindley's equation.

M. Miyazawa, *A Formal Approach to Queueing Processes in the Steady State and their Applications*, J. of Applied Probability, 16, 1979, pp. 332-346.

This is one of the earliest papers in which the RCL technique was introduced.

R. R. Mazumdar, V. Badrinath, F. Guillemin, and C. Rosenberg, *On Pathwise Rate Conservation for a Class of Semi-martingales*, Stoc. Proc. Appl, 47, 1993, pp. 119-130

This paper generalizes the RCL to include processes of unbounded variation and moreover shows that the property is a pathwise property in that even stationarity is not required. There is balance between the derivatives and jumps in an empirical sense.

M. Miyazawa, *Rate Conservation Laws: A Survey*, Queueing Systems Theory Appl., 15, 1994, pp.1-58

This is a very comprehensive survey of RCL applied in many different contexts including processes of unbounded variation and to fluid Palm measures etc.

Guillemin, F. and Mazumdar, R.; *Rate conservation laws for multidimensional processes of bounded variation with applications to priority queueing systems*, Methodology and Computing in Applied Probability, Vol. 6, 2004, pp. 136-159.

This paper provides generalizations of the RCL to deal with multi-dimensional models that occur in more complicated queueing models involving priorities and other service disciplines.

H. Dupuis, B. Hajek, *A simple formula for mean multiplexing delay for independent regenerative sources*, Queueing Systems Theory Appl. 16 (1994), no. 3-4, 195–239

This paper is on the analysis of fluid queues with ON-OFF inputs using regenerative techniques rather than the fluid Palm measure approach.

P. Brémaud, *A Swiss Army formula of Palm calculus*, J. Appl. Probab. 30 (1993), no. 1, 40–51.

This is a seminal paper on the generalization of the RCL idea to both point processes as well as continuous stationary measures (as in the fluid model) as presented in this chapter. It is shown in the paper all Little's formula type of relations as well as the more general $H = \lambda G$ formulae, of which the Pollaczek-Khinchine formula is an archetype, are simple applications of the idea. The author calls it the Swiss Army formula because of its versatility.

CHAPTER 3

Loss Models for Networks

3.1 INTRODUCTION

Circuit-multiplexed networks are a modern generalization of classical circuit-switched telephone models. Circuit-multiplexed networks are distinguished by the fact that each origin-destination pair of nodes in the network is connected by a fixed path called a **route** involving one or more intermediate switches (or routers referred to as nodes) with no buffering available at the nodes. Once a connection request is admitted the packets will follow a pre-determined (perhaps updated) route from the source node at which a connection is requested to the node to the destination node. There is no buffering provided and thus an arriving connection or call requests some bandwidth, which if available, is allocated to that connection during its lifetime. Since there is no queueing, a connection that arrives and cannot find adequate bandwidth at a server is rejected or blocked. Overloads, even temporary, are not included in the basic model.

A typical scenario of a circuit-multiplexed network including the switch and signalling architecture is shown[1] in Figure 3.1.

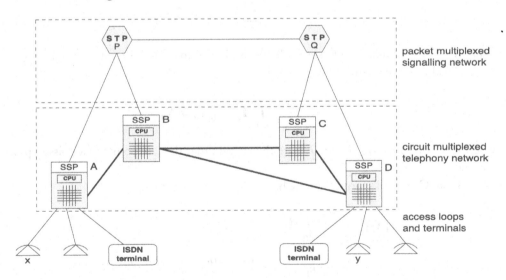

Figure 3.1: A typical circuit-multiplexed network showing switches, access points, signalling network, and trunks (or links).

[1]Reprinted by kind permission of A. Kumar, D. Manjunath, and R. Kuri

Although most network switches or routers have buffers, we will see in the next chapter there is a way of looking at network models in terms of equivalent loss network models. The key performance metric in this context of loss models is the blocking probability, i.e., the probability that an arriving request will not find enough bandwidth on a route to its destination.

In this chapter we will see how to calculate the key performance metric (the blocking probability) for networks by progressively looking at more and more complicated situations. This material can be probed for in greater depth in the books as well as in the papers listed at the end of the chapter.

3.2   MATHEMATICAL PRELIMINARIES

Loss network models are intimately connected to the structure of infinite server queues. Because of their importance in the foregoing modeling and analysis we will see some very important and basic results on $M/G/\infty$ models where arrivals are Poisson with rate λ and each arrival brings an amount of work (the packet length) that is i.i.d. from a distribution $G(.)$. Although they are referred to as queues, there is no queueing involved. An arrival is immediately allocated a server and leaves the system after service. The key performance measure and mathematical description of the state of the system is the number of busy servers at any time t.

Let X_t denote the number of servers that are busy or occupied at time t. Let $\{\sigma_n\}$ denote the service time of each packet and the sequence is i.i.d. with common distribution $G(x)$. Then:

$$X_t = \sum_n \mathbb{1}_{[T_n \leq t < T_n + \sigma_n]} \tag{3.1}$$

Thus, X_t is just the number of arrivals that arrived prior to t but have not yet been served.

An alternative way of writing (3.1) is

$$X_t = \int_0^t \mathbb{1}_{[u \leq t < u + \sigma_u]} dN_u$$

where $\sigma_u = \sigma_n \mathbb{1}_{[T_n \leq u < T_{n+1}]}$ and $\{T_n\}$ are the arrival times that are points of a Poisson process $\{N_t\}$ with rate λ.

From Campbell's formula (see Chapter 1):

$$
\begin{aligned}
\mathbb{E}[X_t] &= \mathbb{E}[\int_0^t \mathbb{1}_{[u \leq t < u + \sigma_u]} dN_u] \tag{3.2}\\
&= \lambda \int_0^t \mathbb{P}(\sigma_0 > t - u) du \ \ (\text{since the } \{\sigma_n\} \text{ are i.i.d.})\\
&= \lambda \int_0^t (1 - G(t - u)) du\\
&= \lambda \int_0^t (1 - G(v)) dv = \lambda \int_0^t \mathbb{P}(\sigma_0 > v) dv \tag{3.3}
\end{aligned}
$$

Therefore, in equilibrium, i.e., as $t \to \infty$, we see

$$\lim_{t \to \infty} \mathbb{E}[X_t] = \lambda \mathbb{E}[\sigma_0] = \rho$$

or the average number of busy servers is just the average load on the system and is finite as long as $\mathbb{E}[\sigma_0] < \infty$.

Thus, the mean only depends on the average service time unlike the $M/G/1$ case where it depends on the second moment of σ_0. This is actually a reflection of an important property called *insensitivity* . Insensitivity means that the stationary distribution depends on the arrival times and the service requirements only through their means rather than the knowledge of their distributions. This is stated and proved below.

Proposition 3.1 *Consider a $M/G/\infty$ system in equilibrium with $\rho = \lambda \mathbb{E}[\sigma_0] < \infty$. Then the stationary distribution of the number of busy servers is completely specified by ρ and has a Poisson distribution given by:*

$$\mathbb{P}(Q_0 = n) = \pi(n) = \frac{\rho^n}{n!} e^{-\rho}, \quad n = 0, 1, \ldots, \tag{3.4}$$

Moreover, in equilibrium the departure process is Poisson with rate λ.

Proof. Assume that $G(.)$ has a density given by $g(x)$. Define :

$$A_t = \{(s, x) \in \mathbb{R} \times \mathbb{R} : s \leq t, s + x > t\}$$

Define the 2-dimensional Poisson process N with intensity $\lambda g(x)$, then the occupation process X_t defined in (3.1) can be written as:

$$X_t = N(A_t)$$

Thus, X_t has a Poisson distribution with parameter

$$\begin{aligned}
\mu(A_t) &= \mathbb{E}[N(A_t)] \\
&= \int_{A_t} \lambda g(x) ds dx \\
&= \lambda \int_0^\infty (t - (t - x)^+) g(x) dx = \lambda \int_0^\infty \min(x, t) g(x) dx
\end{aligned}$$

Hence, $\lim_{t \to \infty} \mu(A_t) = \lambda \int_0^\infty x g(x) dx = \lambda \mathbb{E}[\sigma_0] = \rho$. In other words, as $t \to \infty$ the intensity of the Poisson distribution is a constant and given by (3.4). Since it depends only on ρ we see that the distribution is insensitive to $G(.)$ and only depends on its mean $\mathbb{E}[\sigma_0]$.

The second assertion follows from Proposition 1.16 by noting that the departure times are just random translations of the arrival times by their service requirements, i.e., $d_n = T_n + \sigma_n$. $\square$

Remark 3.2

Consider the $M/M/\infty$ model. This can be modelled as a birth-death process on $\mathbb{Z}+$ with the following transition rates (see Appendix):

$$
\begin{aligned}
q_{i,i+1} &= \lambda \quad i = 0, 1, 2, \cdots \\
q_{i,i-1} &= i\mu \quad i = 1, 2, \cdots ,
\end{aligned}
$$

Now using the fact that the stationary distribution satisfies $\pi Q = 0$, and using the fact that $\sum_j \pi(j) = 1$ we obtain:

$$
\pi(j) = \frac{\rho^j}{j!} e^{-\rho}, \quad j = 0, 1, \cdots , \tag{3.5}
$$

where $\rho = \lambda \mathbb{E}[\sigma] = \frac{\lambda}{\mu}$. Note the stationary distribution exists as long as $\rho < \infty$ (or the service time has finite mean).

It is easy to see from the solution that the $M/M/\infty$ queue is reversible by showing that detailed balance equation (A.6) holds. This also establishes the fact that the departures from the $M/M/\infty$ system form a Poisson process with rate λ.

From reversibility, it follows that the $M/M/C/C$ queue can be viewed as the $M/M/\infty$ queue restricted to the states $\{0, 1, 2, \cdots , C\}$.

Now, in light of the insensitivity of the $M/G/\infty$ system in equilibrium it is easy to check that detailed balance holds since $\lambda_i = \lambda$ and $\mu_i = i\mu$ where $\mu = (\mathbb{E}[\sigma_0])^{-1}$, and thus the $M/G/C/C$ system will have the same stationary distribution as the $M/M/C/C$ system.[2]

3.3   ERLANG LOSS SYSTEM

The Erlang loss model is the name given to an $M/G/C/C$ system with C servers and no waiting room.

In this model calls arrive as a Poisson process with rate λ and the calls last (hold a resource) a random amount of time with mean $\frac{1}{\mu}$. Suppose there are C circuits or lines available. Then when C lines are busy, arriving calls are lost. Assume that the system is in equilibrium.

Let us calculate the probability that an arriving call finds C circuits busy and is thus blocked. This is called the blocking probability.

Let $\pi(j) = \mathbb{P}(j$ circuits are busy$)$. Then $\pi_C(j)$ is given by:

$$
\pi_C(j) = \frac{\pi(j)}{\sum_{n=0}^{C} \pi(n)}, \quad j = 0, 1, \ldots, C \tag{3.6}
$$

---

[2]This part actually needs more proof because the congestion process of the $M/G/\infty$ queue is not Markov by itself. The proof uses the fact that any general distribution can be obtained as the limit of phase-type distributions(of which the Erlang-k is a special case) and one can construct a Markov process by including the phases of the service and then establishing reversibility of that queue. See the book of Kelly on Reversibility and Stochastic Networks listed in the references.

where $\pi(n)$ is given by (3.4).

Now let us see how to calculate the probability that an arriving call is blocked. Define

$$\pi_B = \lim_{T \to \infty} \frac{1}{A(0,T)} \int_0^T \mathbb{1}_{[X_{s-}=C]} dA_s \tag{3.7}$$

where $A(0,t) = A_t$ denotes the number of arrivals in $(0,t)$.

This is the long term fraction of arrivals that find that C lines are busy and corresponds in the limit to the probability that an arrival sees all the C lines busy. Noting that A_t is Poisson we obtain:

$$\pi_B = \pi_C(C) = E_B(\rho, C) = \frac{\rho^C}{C!} \left(\sum_{j=0}^C \frac{\rho^j}{j!} \right)^{-1} \tag{3.8}$$

$E_B(\rho, C)$ is called the Erlang-B formula . Note the apparition of PASTA in equation (3.7) since A_t, the arrival process, is Poisson.

Now let us consider a more complicated system, this results in the so-called Engset formula. The main difference is that in this case there is a finite population of users.

Consider the same problem as before: there are M users each of whom attempts a call at rate λ. The rest is as above. We now want to calculate the probability that a call is blocked.

Once again it is easy to see that the generator of the birth-death process is given by:

$$\begin{aligned} q_{i,i+1} &= (M-i)\lambda, \quad i = 0, 1, 2, \cdots, C-1 \\ q_{i+1,i} &= (i+1)\mu \quad i = 1, 2, \cdots, C-1 \end{aligned}$$

Now from $\pi Q = 0$ we obtain:

$$\pi_M(j) = \pi_M(0) \binom{M}{j} \rho^j, \quad j = 0, 1, \cdots, C$$

and $\pi_M(0)$ is calculated from $\sum_{j=0}^C \pi_M(j) = 1$.

Now, since the arrival rate depends on the state of the system, to obtain the blocking probability, the equivalent of equation (3.7) is now given by $\pi_A(C)$ where:

$$\pi_A(j) = \frac{\lambda(M-j)\pi_M(j)}{\sum_{i=0}^C \pi_M(i)\lambda(M-i)}.$$

Substituting for $\pi_M(j)$ and simplifying we obtain:

$$\pi_A(C) = \frac{\rho^C \binom{M-1}{C}}{\sum_{j=0}^C \rho^j \binom{M-1}{j}} = \pi_{M-1}(C) \tag{3.9}$$

In other words, an arrival *sees* the system busy with $M-1$ users attempting to access the system. This is called the Engset formula . The Engset formula reduces to the Erlang-B formula when the population M goes to infinity in such a way as to keep the attempt rate equal to λ.

This is a classical case when the arrival distribution is different from the stationary distribution. This is because the arrival process is no more Poisson and hence PASTA does not apply.

Below are some properties of the Erlang-B formula .

Proposition 3.3

Let $E_B(\rho, C)$ denote the Erlang-B formula for a system with C servers and traffic intensity ρ. Then:

1. $E_B(\rho, C)$ is increasing in ρ

2. $E_B(\frac{\rho}{C}, 1) \geq E_B(\rho, C)$

3. For $L \in (0, 1)$, let $E_B^{-1}(L)$ be the inverse Erlang-B function for a fixed capacity C, then $E_B^{-1}(L)$ is strictly increasing in $L \in [0, 1]$.

Proof. We first prove 1). Let X_t denote the number of busy servers. Then, by the application of Little's law we obtain:

$$\mathbb{E}[X_t] = \lambda(1 - E_B(\rho, C))\mathbb{E}[\sigma_0] = \rho(1 - E_B(C, \rho))$$

Noting that $\mathbb{E}[X_t] \leq C$ we obtain:

$$E_B(\rho, C) \geq (1 - \frac{C}{\rho})^+$$

Now from the definition of $E_B(\rho, C)$ we obtain:

$$\frac{\partial E_B(\rho, C)}{\partial \rho} = E_B(\rho, C)[\frac{C}{\rho} - 1 + E_B(\rho, C)]$$

From the previous step we obtain that the expression in the brackets is always non-negative and this establishes that $E_B(\rho, C)$ is non-decreasing in ρ.

The proof of 2) follows from the fact that we can write:

$$E_B(\rho, C) = \frac{\frac{\rho}{C} E_B(\rho, C - 1]}{1 + \frac{\rho}{C} E_B(\rho, C - 1)}$$

and noting that $E_B(\rho, C - 1) \leq 1$.

The proof of part 3) follows from part 1 since by definition, $E_B^{-1}(L)$ corresponds to the value of *rho* for which $E_B(\rho, C) = L$, and from 1) it is increasing in ρ. □

The quantity $E_B(\frac{\rho}{C}, 1)$ can be interpreted as the probability that an arrival following a blocked arrival is also blocked assuming the services are exponential. To see this: The probability that an arrival following a blocked arrival is also blocked is equal to the probability that an arrival takes place before a call departs given that all the C servers are busy. This probability is just $\frac{\lambda}{\lambda + C\mu} = E_B(\frac{\rho}{C}, 1)$.

We now extend our study to models that have more relevance to today's networks. These are the so-called multi-rate Erlang models. They arise when different arrivals require a different number of circuits (multiple number of circuits or units of bandwidth)- or differing amount of bandwidth.

3.4 MULTI-RATE ERLANG LOSS SYSTEMS

The multi-rate loss model is necessary when we wish to model the modern circuit-switched networks where arriving calls or sessions can have differing bandwidth requirements.

Let us begin with the direct generalization of the classical Erlang loss model. Specifically, consider the following model:

- There are M types of calls, connections, or sessions.

- A call of class or type k, $k = 1, 2, \cdots, M$ arrives as a Poisson process with rate λ_k.

- Class k holds the resource or line for a mean of $\mathbb{E}[\sigma_k]$ units of time.

- A call of class k requires A_k units of bandwidth or circuits.

- Different classes are independent.

- On completion all resources are simultaneously released.

- The resource or switch has a total of C circuits or units of bandwidth available.

- An arriving call of class k that finds less than A_k units of bandwidth available is blocked and lost.

Let $\mathbf{n} = (n_1, n_2, \cdots, n_M)$ be the state of the system that denotes the number of types of calls of each type in progress. Let S denote the state space given by:

$$S = \{\mathbf{n} : \sum_{k=1}^{M} n_k A_k \le C\}$$

Define $\rho_k = \lambda_k \mathbb{E}[\sigma_k]$, $= 1, 2, \cdots, M$. Then, we can state the following result.

Proposition 3.4

Let $\mathbf{n}$ be the vector of the number of calls at each time in progress in the system at a given time. Then, in equilibrium:

$$\pi(\mathbf{n}) = \frac{1}{G(C, M)} \prod_{k=1}^{M} \frac{\rho_k^{n_k}}{n_k!} \tag{3.10}$$

where

$$G(C, M) = \sum_{\{\mathbf{n}: A\mathbf{n} \le C\}} \prod_{k=1}^{M} \frac{\rho_k^{n_k}}{n_k!} \tag{3.11}$$

and $A\mathbf{n} \le C \Leftrightarrow \sum_{k=1}^{M} n_k A_k \le C$.

Proof. This result follows because of the truncation argument for reversible Markov chains. First assume that $C = \infty$, i.e., no capacity limit and the services are exponential.

Then, the system can be modelled as a multi-dimensional birth-death process. Let $\mathbf{e_i} = (0, 0, \cdots, 0, 1, 0, \cdots)$ where the 1 is in the ith place. Then the transitions of this process follow:

$$(\mathbf{n}) \rightarrow (\mathbf{n} + \mathbf{e_j}) \qquad \text{with rate } \lambda_j$$

$$(\mathbf{n}) \rightarrow (\mathbf{n} - \mathbf{e_j}) \qquad \text{with rate } \mu_j n_j$$

where $\mu_j = \frac{1}{\mathbb{E}[\sigma_j]}$.

Hence, it can readily be seen from the characterization of the stationary distribution of the stationary distribution for the $M/M/\infty$ case and because the processes are independent, the Markov chain describing the evolution of the M independent birth-death processes is reversible with stationary distribution denoted by $\tilde{\pi}(\mathbf{n})$ and given by:

$$\tilde{\pi}(\mathbf{n}) = \prod_{k=1}^{M} \frac{\rho_k^{n_k}}{n_k!} e^{-\rho_k}$$

Now since the set $\{\mathbf{n} : A\mathbf{n} \leq C\}$ is closed the result follows by applying the result for reversible processes with truncated state-spaces, i.e.,

$$\pi(\mathbf{n}) = \frac{\tilde{\pi}(\mathbf{n})}{\sum_{A\mathbf{m} \leq C} \tilde{\pi}(\mathbf{m})}$$

Then the passage to general service time distributions follows because of the argument that $M/G/\infty$ and $M/M/\infty$ queues have the same stationary distribution (the insensitivity property).

Now, the quantity of interest to us is the blocking probability that in this case is the probability that an arrival of type k cannot find at least A_k units of bandwidth available.

Define: $S_k = \{\mathbf{n} : C - A_k < \sum_{j=1}^{M} n_j A_j \leq C\}$. Then S_k denotes the set of *blocking states* when there is less than A_k units of bandwidth available.

Hence:

$$\mathbb{P}_{B_k} = \sum_{j \in S_k} \pi(j) = 1 - \sum_{j \in S_k^c} \pi(j)$$

Now noting that $\sum_{j \in S_k^c} \pi(j)$ is just the probability that the system has a capacity (or total available bandwidth) of $C - A_k$, we can write the class-k blocking probability as:

$$\mathbb{P}_{B_k} = 1 - \frac{G(C - A_k, M)}{G(C, M)} \tag{3.12}$$

$\square$

One might think the problem is solved. However, in real systems M could be of the order of 20 and C is typically 10^6. Computing $G(C, M)$ and $G(C - A_k, M)$ is very cumbersome because S has many states. For example, the typical state space is of the order $O(C^M)$. This can be extremely large when $C = 10^6$ and M is 20. The complexity is referred to as $\#P$.

There is a way of simplifying the calculation of the blocking probability for each class by viewing the system in terms of the total server occupancy. This gives rise to a very simple recursion for the server occupancy probability distribution from which the class blocking probabilities can be calculated as is given by equation (3.12) . The way to do this was independently discovered in the 1980s by J. Kaufman and J.R. Roberts although the antecedents go back to earlier times. It is called the Kaufman-Roberts algorithm which is presented below.

The Kaufman-Roberts algorithm is based on viewing the system in terms of the total bandwidth or circuit occupancy rather than the vector of states as we considered above.

Let $q(j) = \mathbb{P}\{j$ units of bandwidth occupied$\}$. Then

$$
\begin{aligned}
q(j) &= \sum_{\{n:An=j\}} \pi(\mathbf{n}) \\
&= \sum_{\{n:An=j\}} \frac{\rho_i^{n_i}}{n_i!} \frac{1}{G(C, M)}
\end{aligned}
$$

Then we can write the blocking probability for class k as:

$$
\mathbb{P}_{B_k} = \sum_{i=0}^{A_k-1} q(C - i), \quad k = 1, 2, \cdots, M
$$

The interpretation is that an arrival of class k is blocked because there are less than A_k units of bandwidth available and so the server occupancy is either in $C, C - 1, \cdots, C - (A_k - 1)$. The Kaufman-Roberts algorithm is then based on the following result:

Lemma 3.5 *Let $q(i)$ be defined as above. Then:*

$$
\sum_{i=1}^{M} \rho_i q(j - A_i) A_i = j q(j), \quad j = 1, 2, \cdots, C \tag{3.13}
$$

where $q(j) = 0$, $j < 0$ and $\sum_{j=0}^{C} q(j) = 1$

Proof. Define $S(j) = \{\mathbf{n} : A\mathbf{n} = j\}$ and $R_k(j) = \sum_{\{\mathbf{n} \in S(j)\}} n_k \pi(\mathbf{n})$ Then:

$$
\begin{aligned}
jq(j) &= \sum_{\mathbf{n} \in S(j)} j\pi(\mathbf{n}) \\
&= \sum_{\mathbf{n} \in S(j)} [\sum_{k=1}^{M} A_k n_k] \pi(\mathbf{n}) \\
&= \sum_{k=1}^{M} A_k \sum_{\mathbf{n} \in S(j)} n_k \pi(\mathbf{n}) \\
&= \sum_{k=1}^{M} A_k R_k(j)
\end{aligned}
$$

From the definition of $\pi(\mathbf{n})$ we have for $n_k > 0$:

$$
\begin{aligned}
n_k \pi(n) &= \frac{n_k}{G(C,M)} \prod_{j=1}^{M} \frac{\rho_j^{n_j}}{n_j!} \\
&= \frac{\rho_k^{n_k}}{G(C,M)(n_k-1)!} \prod_{j \neq k} \frac{\rho_j^{n_j}}{n_j!} \\
&= \rho_k \pi(\mathbf{n} - e_k)
\end{aligned}
$$

Thus, if $n_k > 0$,

$$
\begin{aligned}
R_k(j) &= \sum_{\mathbf{n} \in S(j)} n_k \pi(\mathbf{n}) \\
&= \rho_k \sum_{\mathbf{n} \in S(j)} \pi(\mathbf{n} - e_k) \\
&= \rho_k \sum_{\mathbf{n} \in S(j - A_k)} \pi(\mathbf{n}) \\
&= \rho_k q(j - A_k)
\end{aligned}
$$

Hence, the result follows.   □

Noting that $\mathbb{P}_{B_k} = \sum_{j=0}^{A_k-1} q(C-j)$ we now give the Kaufman-Roberts algorithm.

Kaufman-Roberts Algorithm

1. Set $g(0) = 0$ and $g(j) = 0$, $j < 0$

2. For $j = 1, 2, \cdots, C$, set

$$g(j) = \frac{1}{j} \sum_{k=1}^{M} A_k \rho_k g(j - A_k)$$

3. $G = \sum_{j=1}^{C} g(j)$

4. For $j = 0, 1, \cdots, C$, set $q(j) = \frac{g(j)}{G}$

5. For $k = 1, 2, \cdots, M$ set $\mathbb{P}_{B_k} = \sum_{j=C-A_k+1}^{C} q(j)$

The complexity of this algorithm is $O(CM)$ which is much more efficient than $O(C^M)$. Finally, the server utilization is given by $\sum_{j=1}^{C} j q(j)$. This represents the average number of servers busy or the average amount of bandwidth used.

The multi-rate Erlang blocking formula does not necessarily imply monotonicity in blocking if the load of a given class increases. What can be shown is that the class whose load increases has its blocking increased but the blocking of a class that has a smaller A_k can decrease. The reference of Nain listed at the end of the chapter discusses the monotonicity of blocking probabilities for multi-rate Erlang loss systems and conditions when it is true.

3.5   THE GENERAL NETWORK CASE

So far we have concentrated our focus on single-link systems. We now discuss the network case. To understand the network case let us first define a few quantities.

- J links (either physical or logical)

- A capacity of C_j is available on link $j \in \{1, 2, \cdots, J\}$.

- There are R routes linking source-destination pairs. Typically, $r \in R$ implies $r = \{j_1, j_2, \ldots, j_r\}$. In physical networks $R \geq J$.

- Calls are specified by routes and on a given link j belonging to route r the call uses $A_{j,r}$ units of bandwidth.

- Holding times random with unit mean without loss of generality.

- A connection request holds all links simultaneously and releases them simultaneously.

- Arrivals of requests for route $r \in R$ are Poisson rate ν_r.

Then it can be shown that in equilibrium the stationary distribution of the number of connections along each route is given by:

Proposition 3.6

Let $A = \{A_{j,r}\}_{j,r=1}^{J,M}$ be the so-called incidence matrix or bandwidth matrix with the convention that $A_{j,r} = 0$ if class r specified by the route does not use link j.

Let $\mathbf{n} = (n_1, n_2, \ldots, n_R)$ be the number of connections of each route in the network. Then the following result holds:

$$P(\mathbf{n}) = G(C)^{-1} \prod_{r \in R} \frac{v_r^{n_r}}{n_r!} \qquad \mathbf{n} \in S(C)$$

where

$$S(C) = \{\mathbf{n} : A\mathbf{n} \leq C\}$$

and

$$G(C) = \sum_{\mathbf{n} \in S(C)} \prod_{r \in R} \frac{v_r^{n_r}}{n_r!} .$$

The proof follows from the properties of the $M/G/\infty$ system. This is because in equilibrium, the departures from an $M/G/\infty$ system are Poisson and thus independence between classes (now defined by the routes) is maintained. Then one uses the fact that the underlying Markov chain is reversible in equilibrium and so the restriction to the truncated state-space now specified by the link capacities and follows from Proposition A.22.

To obtain the blocking probability we have to define the blocking states. The quantity of interest is the blocking along a given route $r \in R$ which we denote by $\mathbb{P}_{B_r}$. Define:

$$M_{j,r} = \{\mathbf{n} \in S : \sum_s A_{j,s} n_s \leq C_j - A_{j,r}\}$$

Then $M_{j,r}$ denotes those states on link j where there is at least $A_{j,r}$ units of bandwidth or circuits available.

Similarly, one can define :

$$\mathcal{K} = \{\mathbf{n} : \sum_r A_{j,r} n_r = k_j\}$$

which denotes the set of states when k_j is the units occupied on link j and

$$\mathcal{K}_{j,r} = \{\mathbf{k} : C_j - A_{j,r} < k_j \leq C_j\}$$

Now define $M_r = \bigcap_j M_{j,r}$ and $\mathcal{K}_r = \bigcup_j \mathcal{K}_{j,r}$.
Then :

$$
\begin{aligned}
\mathbb{P}_{B_r} &= 1 - \sum_{M_r} P(\mathbf{n}) && (3.14) \\
&= q(\mathcal{K}_{1,r} \cup \ldots \cup \mathcal{K}_{J,r}). && (3.15)
\end{aligned}
$$

where $q(k_j) = \mathbb{P}(n : \sum_r A_{j,r} n_r = k_j)$ denotes the probability that k_j units of bandwidth are occupied on link j.

Of course the formulae for computing the blocking probabilities are now even more computationally demanding than in the single link case. It is illustrative to see how this can be applied by considering the following example of three links.

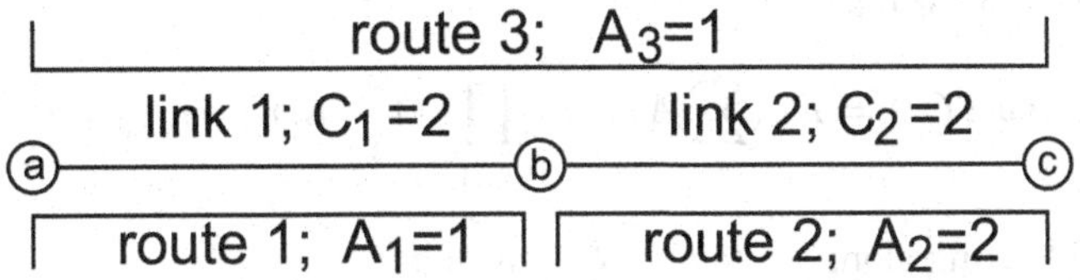

Figure 3.2: Three route network with 2 links.

In this example there are three classes of traffic and 2 links. The routes are $\{(a, b), (b, c), (a, c)\}$ and Class 1 corresponds to route (a, b), Class 2 corresponds to (b, c) and Class 3 corresponds to (a, c). Let $S = \{s_k\}$ denote the state space where $s_k = (n_1, n_2, n_3)$ where n_i are the number of connections of Class i.

In this simple example we can evaluate all the distinct states :

$$S = \{(0, 0, 0), (1, 0, 0), (1, 1, 0), (0, 1, 0), (1, 0, 1), (0, 0, 1), (0, 0, 2), (2, 0, 0), (2, 1, 0)\}$$

and we can also find the blocking states for each class:

$$
\begin{aligned}
S_1 &= \{(2, 0, 0), (0, 0, 2), (2, 1, 0)\} \\
S_2 &= \{(0, 0, 1), (1, 0, 1), (0, 1, 0), (1, 1, 0), (0, 0, 2), (2, 1, 0)\} \\
S_3 &= \{(0, 1, 0), (1, 1, 0), (2, 0, 0), (0, 0, 2)\}
\end{aligned}
$$

Now, the respective blocking probabilities are obtained by summing over blocking states:

$$
\begin{aligned}
\mathbb{P}_{B_1} &= \pi(2, 0, 0) + \pi(0, 0, 2) + \pi(2, 1, 0) \\
\mathbb{P}_{B_2} &= \pi(0, 0, 1) + \pi(1, 0, 1) + \pi(0, 1, 0) + \pi(1, 1, 0) + \pi(0, 0, 2) + \pi(2, 1, 0) \\
\mathbb{P}_{B_3} &= \pi(0, 1, 0) + \pi(1, 1, 0) + \pi(2, 0, 0) + \pi(0, 0, 2)
\end{aligned}
$$

and we note: $\pi(n_1, n_2, n_3) = G^{-1} \prod_{i=1}^{3} \frac{v_i^{n_i}}{n_i!}$ and $G = \sum_{n \in S} \pi(n)$

In general, evaluating all the blocking states might be very cumbersome especially if the network state has a very high dimension and the number of links is large. However, there are some approximations possible if we assume that the blocking to be expected on a given link is small. This leads to the so-called Erlang fixed point ideas that were presented by Kelly in his seminal work.

Let us first develop some intuition behind the results. Consider the classical Erlang loss model that corresponds to the single rate case and where $A_{j,r} = 1$ or 0 and $A_{j,r} = 1$ if route r

uses link j. Assume the blocking along each link is independent of the others. Let L_j denote the probability of being blocked in link j. Then the total load arriving on link j with capacity C_j is just $\rho_j = \sum_r A_{j,r}\rho_r \prod_{i\in r-\{j\}}(1-L_i)$. We should only consider the upstream links along the route r in computing the product but if L_i is small we consider the product over all links on the route.

Now by independence, and the Poisson assumption of arrivals to the network we can assume that the aggregate arriving flow on link j is Poisson. Hence, we can compute the blocking on that link using the Erlang formula, i.e.,

$$L_j = E_B(\rho_j, C_j) = E_B\left(\sum_r A_{j,r}\rho_r \prod_{i\in r-\{j\}}(1-L_i), C_j\right), \quad j = 1, 2, \cdots, J \qquad (3.16)$$

And hence the blocking along route $r \in R$ is given by:

$$P_{B_r} = 1 - \prod_{j\in r}(1-L_j) = 1 - \prod_j (1-L_j)^{A_{j,r}} \qquad (3.17)$$

We notice that the calculation of L_j involves knowledge of the other $L_i's$ along the route. Thus, the natural question is how can we compute the solution and indeed, does a solution exist to the set of non-linear equations. The existence and uniqueness was shown by Kelly. We state the result below.

Proposition 3.7 *For loss networks in which arrivals require 1 unit of bandwidth along all links and are only differentiated by their arrival rates, there exists a unique solution to the set of non-linear equations given by (3.16) referred to as the Erlang fixed point.*

Proof. The proof essentially follows from the Brouwer fixed point theorem by the following observation. Let $\mathbf{L} = col(L_1, L_2, \cdots, L_j)$ then one can write $\prod_{i\in r, i\neq j}(1-L_i)$ as $(1-L_j)^{-1}\prod_{i\in r}(1-L_i)$ and thus writing it in vector format: $\mathbf{L} = f(\rho, \mathbf{L})$ where $f(.) : [0,1]^J \rightarrow [0,1]^J$ and $f_j(\rho, \mathbf{L}) = E_B(\rho, \mathbf{L}, C_j)$ the Erlang loss function for link j with the offered load as discussed. Thus, since $f(.)$ is continuous and maps compact region $[0,1]^J$ to $[0,1]^J$ which is compact, by Brouwer's fixed point theorem there exists a fixed point.

The proof of uniqueness is more complicated. Essentially it follows that the fixed point is the stationary point of a convex optimization problem that is shown below.

Let $E_{B_j}^{-1}(L)$ correspond to the inverse of the Erlang-B function for a link of capacity C_j. From the fact that $E_{B_j}^{-1}(L)$ is increasing in $L \in [0,1]$ we have that $\int_0^L E_{B_j}^{-1}(z)dz$ is strictly convex.
Define:

$$F(L) = \sum_{j=1}^J \sum_r A_{j,r}\rho_r \prod_{i\in r}(1-L_i) + \sum_{j=1}^J \int_0^{L_j} E_{B_j}^{-1}(u)du$$

Then, the Hessian $\{H_{i,j}\}$ of $F(L)$ is strictly positive definite with diagonal $H_{j,j} = \frac{d}{dL_j}E_{B_j}^{-1}(L_j) > 0$ and off-diagonal terms $H_{i,j} = \sum_r A_{i,r}A_{jr}\prod_{k\neq i,j}(1-L_k) > 0$ implying that

$F(L)$ is strictly convex in $[0, 1]^J$. Now, the fixed point equation (3.16) just corresponds to the equation obtained for the optimum $\frac{\partial F(L)}{\partial L_j} = 0$ which is unique because $F(L)$ is convex. This completes the proof. □

There are two natural questions associated with the Erlang fixed point problem. They are:1) Is there a simple computational algorithm to find the fixed points? 2) Since the basis for the equivalent load on a link is the independent thinning along the route, how accurate is the result?

We will answer 2) first. The basic result that was shown by Kelly was that the result is asymptotically correct, i.e., when the link capacities are large in comparison to the required bandwidths of each connection (here 1), then the result is very accurate. We make this more precise below.

Consider a *large* network where the link capacities C_j are actually scaled according to a scaling parameter N, i.e., $C_j = C_j(N)$ such that $\lim_{N\to\infty} \frac{C_j(N)}{N} \to C_j$. Also assume the arrival rates are $\lambda_r(N)$ where once again $\frac{\lambda_r(N)}{N} \to \lambda_r$ as $N \to \infty$. Hence, $\rho_r(N) = \lambda_r(N)\mathbb{E}[\sigma_r] \to \rho_r$. The larger the value of N, the capacities and arrival rates scale in such a manner that the limiting system corresponds to the situation where the ratio of arrival rates and capacities are the same as the original system of interest. Let $B_r(N)$ be the blocking probability on route r in the scaled system where $L_j(N)$ denotes the blocking on link j..

The result of Kelly is that $B_r(N) \to B_r$ where:

$$B_r = 1 - \prod_{j \in r}(1 - L_j)^{A_{j,r}}$$

where the L_j are obtained as solutions to the Erlang fixed point of the system with link capacities C_j and arrival rates λ_r.

Let us now return to the problem of computing the fixed-point. The standard method for computing fixed points of functions that are contractions is to perform the iterations:

$$X_{n+1} = f(X_n)$$

If $f(.)$ is a contraction then $X_n \to x$ where $x = f(x)$. Here there is one issue , successive applications of the blocking formula does not generate a monotone sequence of iterates. This follows from the fact that the Erlang-B function is not a contraction (as a function of ρ). Let $\rho_j(L) = \frac{1}{(1-L_j)} \prod_{i \neq j}(1 - L_i)^{A_{i,r}}$. Define $\rho(L) = col(\rho_j(L)$. Then, from the monotonicity of the Erlang-B function, $E_B(\rho(L^1), C) \geq E_B(\rho(L^2), C)$ whenever $L^2 > L^1$. Then define $E_B^2(\rho_1(L), C) = E_B(\rho(E_B(\rho(L), C), C)$. Then, at the first iteration solutions obtained get interchanged or $E_B^2(\rho(L^1, C) \leq E^2(\rho(L^2), C)$, and so on.

Hence, consider the following scheme:. Let $\rho(L^k)$ denote the vector $\rho(L)$ at the $k - th$ iteration.

Define: $L^0 = 0$ (all elements of the vector are 0), $L^1 = 1$ (all elements of the vector are 1). Generate the even iterates L^{2k} and the odd iterates L^{2k+1} starting with L^0 and L^1 respectively. Then it is easy to see because of the order inverting property described above:

$$0 = L^0 < L^2 < \cdots < L^{2k} < \cdots < L^* < \dots < L^{2k+1} < \cdots L^1 = 1$$

i.e., the even iterates starting from 0 will form a monotone increasing sequence and the odd iterates will form a monotone decreasing sequence starting from 1.

Hence, $\lim L^{2k} \leq L^* \leq \lim L^{2k+1}$. In the case we consider when $A_{j,r} = 1$ for all $j \in r$, by uniqueness of the fixed points the two limits will coincide and be the Erlang fixed point. The proof of uniqueness when the $A_{j,r} \neq 1$ is not known and in that case the two iterates may have different limits. However, in most large networks, uniqueness holds due to Kelly's result.

In the multi-rate case there are no general results. Part of the difficulty is that the multi-rate Erlang blocking formula does not possess any monotonicity properties as in the single rate case. One situation that can be treated is when the bandwidth requirements of class k are A_k along each link on the route. Let $L_{j,k}$ denote the blocking on link j for class k, i.e., less than A_k units of bandwidth available on that link. Assume links are independent.

Use the link thinning to compute

$$L_{j,k} = Q_k(C_j, \sum_{l \in K_j} \rho_l \prod_{i \in r-\{j\}} (1 - L_{i,l}))$$

where $Q_k(C, \rho_l) = 1 - \sum_{l=0}^{C-A_k} q(l)$ and $q(l) = \mathbb{P}(l$ servers are occupied$)$ then one can show if the link capacities C_j are large the iterations will converge. Convergence of iterative schemes for the general situation are not known.

It turns out that when systems are large we can obtain explicit $O(1)$ complexity formulae for blocking with an error term that goes to 0 as the relative size (or capacity) of the system to individual bandwidth requirements goes to infinity. This is extremely useful for dimensioning practical systems where link bandwidths are much larger than individual source bandwidth requirements. We will address this issue in the next section beginning with the single-link case.

3.6   LARGE LOSS SYSTEMS

Large loss systems are systems where the number of available servers is large compared to the number of circuits or bandwidth required by a single call or connection. We begin by considering a single resource loss system. The technique that will be used can be generalized to the case of large loss networks.

The model of the large loss system is almost identical to the multi-rate Erlang loss model we considered earlier:

- There are M types of calls, connections, or sessions.

- A call of class or type k, $k = 1, 2, \cdots, M$ arrives as a Poisson process with rate $N\lambda_k$.

- Class k holds the resource or line for a mean of $\mathbb{E}[\sigma_k]$ units of time.

- A call of class k requires A_k units of bandwidth or circuits.

- Different classes are independent.

- On completion all resources are simultaneously released.

- The resource or switch has a total of $NC = C(N)$ circuits or units of bandwidh available.

- An arriving call of class k that finds less than A_k units of bandwidth available is blocked and lost.

The only difference is that both the capacity and rates of arrivals of connections are scaled by N. Thus, the average traffic load on the server is $\rho(N) = \sum_{k=1}^{M} N\lambda_k \mathbb{E}[\sigma_k] A_k = N \sum_{k=1}^{M} \rho_k A_k$ and the ratio of $\rho(N)$ to $C(N)$ is $\frac{\rho}{C}$ where the unscaled quantities can be thought of as quantities related to a *nominal system*. Note also that the ratio of $C(N)$ to A_k is $O(N)$—in other words the large system is an N-fold scaling of the nominal system.

Let $\pi_N(\mathbf{n})$ denote the stationary distribution of the system.

Then:

$$\pi_N(\mathbf{n}) = \frac{1}{G(NC, M)} \prod_{k=1}^{M} \frac{(N\rho_k)^{n_k}}{n_k!}, \qquad (3.18)$$

where

$$G(NC, M) = \sum_{\{\mathbf{n}: A\mathbf{n} \leq NC\}} \prod_{k=1}^{M} \frac{(N\rho_k)^{n_k}}{n_k!}, \qquad (3.19)$$

and $A\mathbf{n} \leq NC \Leftrightarrow \sum_{k=1}^{M} n_k A_k \leq NC$.

The only change in the formula over the multi-rate Erlang formula that we saw before is the normalizing factor $G(NC, M)$ and ρ_k being replaced by $N\rho_k$.

We now state the main result for the blocking of class k connections and indicate the main ideas of the proof that relies on so-called local limit, large deviations theorems due to Bahadur-Rao and Petrov using an exponential twisting idea that allows to find a new probability distribution that is absolutely continuous with respect to the original under which we can set the mean to a value we choose. These are given in the Appendix.

Let $\mathbb{P}_k(N)$ denote the blocking probability of class k in the scaled system. Define:

$$S_k(N) = \{\mathbf{n} : NC - A_k < \sum_{j=1}^{M} n_j A_j \leq NC\}$$

Then $S_k(N)$ denotes the set of *blocking states* when there is less than A_k units free and defines the blocking states. Then let $\mathbb{P}_k(N)$ denote the blocking probability for class k and is given by:

$$\mathbb{P}_k(N) = \sum_{j \in S_k(N)} \pi(j) = 1 - \sum_{j \in S_k^c(N)} \pi(j)$$

We consider the following 3 cases or regimes because the corresponding expressions for the blocking probability are different:

(**Light Load**) $\sum_1^M \lambda_k A_k < C$

(**Critical load**) $\sum_1^M \lambda_k A_k = C$

(**Heavy load**) $\sum_1^M \lambda_k A_k > C$

Proposition 3.8   *Consider the large multi-rate system described above. Then the following estimates for the blocking probability of class k denoted by $\mathbb{P}_k(N), k = 1, 2, \ldots, M$ hold for the Light, Critical and Heavy load regimes:*

Light load regime

$$\mathbb{P}_k(N) = \exp(\tau_C d\epsilon)\frac{\exp(-NI(C))(1 - \exp(\tau_C A_k))}{\sqrt{2\pi N}\sigma(1 - \exp(\tau_c d))}\left(1 + O(\frac{1}{N})\right) \qquad (3.20)$$

Critical load case

$$\mathbb{P}_k(N) = \sqrt{\frac{2}{\pi N}}\frac{A_k}{\sigma}\left(1 + O(\frac{1}{\sqrt{N}})\right) \qquad (3.21)$$

Heavy load regime

$$\mathbb{P}_k(N) = (1 - \exp(\tau_C A_k))(1 + O(\frac{1}{N})) \qquad (3.22)$$

The parameters $I(C), \tau_C, \epsilon, \sigma, \delta$ and d are defined as

- d *is the GCD of* $\{A_1, A_2, \ldots, A_M\}$
- $\epsilon = \frac{NC}{d} - \lfloor(\frac{NC}{d})\rfloor$
- τ_C *is the unique solution to* $\sum_1^M \lambda_k A_k \exp(\tau_C A_k) = C$
- $I(C) = C\tau_C - \sum_1^M \lambda_k(\exp(\tau_C A_k) - 1)$
- $\sigma^2 = \sum_1^M \lambda_k A_k^2 \exp(\tau_C A_k)$

Proof. (Outline)
      We only provide the outline of the main arguments. The proof of the result hinges on the following facts:

Infinite divisibility of Poisson measures

      Let $\xi_k(N)$ be Poisson r.v.'s with mean $N\rho_k$ and $\eta = \sum_{k=1}^M A_k\xi_k$ where ξ_k are independent copies of Poisson r.v.'s with mean ρ_k. Then $\sum_1^N \eta_i$ where η_i are independent copies of r.v.'s with the same distribution as η, and $\sum_{k=1}^M A_k\xi_k(N)$ have the same distribution.

Exponential centering or twisting of distributions

Let $\phi(t)$ be the moment generating function of a distribution $\mu(dx)$. Suppose that $\exists \tau_a$ such that $\frac{\phi'(\tau_a)}{\phi(\tau_a)} = a$. Then define the new distribution

$$\mu_a(dx) = \frac{\exp(\tau_a x)}{\phi(\tau_a)} \mu(dx)$$

Then under μ_a the r.v. has mean a and variance $\sigma^2 = \frac{\phi''(\tau_a)}{\phi(\tau_a)} - a^2$.

Now define $I(a) = \sup_t\{at - \log(\phi(t))\}$, $I(a)$ is called the rate function. Let μ^{*n} denote n fold convolution of μ, then

$$\mu^{*n}(dx) = \exp(-nI(a)) \exp(\tau_a(na - x))\mu_a^{*n}$$

Local Gaussian Approximations

We take μ corresponding to the distribution of η and $a = C$ and then use the fact that the n-fold convolution of a measure can be approximated by a Gaussian distribution (Local limit theorem) with variance given earlier.

We use these ideas as follows: if we neglect the denominator of (3.18) and multiply each term in the product by $e^{-N\rho_k}$, then the numerator can be viewed as distribution independent Poisson r.v.'s. $\mathbf{n} = (n_1, n_2, \ldots, n_M)$ with n_k having mean $N\rho_k$. Let us denote the joint distribution $\hat{\pi}(\mathbf{n})$. Now summing $\hat{\pi}(\mathbf{n})$ over $S_k(N)$, the blocking states for class k we see that the states are small perturbations (of max size A_k) around NC and thus if we use the exponential centering to center the r.v.'s around NC then we can use a local Gaussian approximation to approximate $\mu_{NC}^*(x)$ where $x \le A_k$. We then use the interpretation that the denominator $G(NC, M)$ is just the probability that the sum of the M r.v.'s $(n_1, n_2, \ldots, n_M)$ weighted by their bandwidth requirements is less than NC. In the light load case this is close to 1 being the measure of the r.v.'s whose sum has mean NC under μ_{NC}. Details can be found in the paper by Gazdicki, Likhanov, and Mazumdar listed at the end of the chapter. $\square$

A natural question is how accurate are these methods. Below are some numerical results that were generated using the closed form formulae given by (3.20), (3.21), and (3.22). It is clear the accuracy improves with the size of the system but is quite accurate even for moderate sized systems.

In the tables below we give some numerical evidence of the accuracy of the asymptotic approach. We compare the asymptotic results with the exact results obtained by running the recursive algorithm due to Kaufman-Roberts algorithm.

The calculations were performed on a loss system with two classes of traffic. The capacity C was assumed to be 20. Traffic of class 1, whose blocking probability is denoted by P_1, was assumed to require 2 circuits. Traffic of class 2, whose blocking probability is denoted by P_2, was assumed to require 5 circuits, i.e., $A_1 = 2$ and $A_2 = 5$. The different cases correspond to variations in their traffic intensities which are denoted by λ_1 and λ_2 respectively.

Table 3.1: LIGHT LOAD : $\lambda_1 = 1.0, \lambda_2 = 2.0$

Scaling	Exact			Approximate	
N	$P_1(N)$	$P_2(N)$		$P_1(N)$	$P_2(N)$
1	0.768 E-01	0.176 E+00		0.540 E-01	0.161 E+00
5	0.347 E-02	0.104 E-01		0.343 E-02	0.103 E-01
10	0.212 E-03	0.634 E-03		0.212 E-03	0.634 E-03
15	0.151 E-04	0.452 E-04		0.151 E-04	0.452 E-04
20	0.114 E-05	0.342 E-05		0.114 E-05	0.342 E-05
25	0.893 E-07	0.267 E-06		0.893 E-07	0.267 E-06

Table 3.2: CRITICAL LOAD : $\lambda_1 = 5.0, \lambda_2 = 2.0$

Scaling	Exact			Approximate	
N	$P_1(N)$	$P_2(N)$		$P_1(N)$	$P_2(N)$
1	0.169 E+00	0.428 E+00		0.191 E+00	0.477 E+00
5	0.811 E-01	0.203 E+00		0.853 E-01	0.213 E+00
10	0.582 E-01	0.146 E+00		0.603 E-01	0.151 E+00
15	0.478 E-01	0.120 E+00		0.492 E-01	0.123 E+00
20	0.416 E-01	0.104 E+00		0.426 E-01	0.107 E+00
25	0.373 E-01	0.932 E-01		0.381 E-01	0.954 E-01
101	0.188 E-01	0.469 E-01		0.190 E-01	0.474 E-01
191	0.137 E-01	0.342 E-01		0.138 E-01	0.345 E-01

This approach for single resource loss systems can be extended to the network case also using the same ideas above. Indeed, in the large loss network case we can obtain explicit results in all regimes provided they hold uniformly on all links within the network. This is particularly noteworthy because the fixed point techniques are not guaranteed to converge.

The loss network model is as before.

Consider a network with J links and offered calls transfered via R fixed routes. Calls on route $r \in \{1, \ldots, R\}$ occur as a Poisson process with rate $N\nu_r$ and their mean holding time is taken equal to unity. Flows associated with each route are mutually independent. The total number of circuits on link $j \in \{1, \ldots, J\}$ is denoted by NC_j. A call pertaining to route r requires $A_{j,r}$ circuits on link j (with $A_{j,r} = 0$ if $j \notin r$). Without loss of generality, the greatest common divisor d_j of the numbers $A_{j,r}, j \in r$, is set to 1 (if $d_j \neq 1, C_j$ and $A_{j,r}$ can be replaced by $\lfloor C_j/d_j \rfloor$ and $\lfloor A_{j,r}/d_j \rfloor$, respectively). In the following, $\mathbf{A}$ denotes the matrix with elements $A_{j,r}, 1 \leq j \leq J, 1 \leq r \leq R$. Here N denotes the scaling factor that reflects the size of the system.

Table 3.3: HEAVY LOAD : $\lambda_1 = 2.0, \lambda_2 = 5.0$

Scaling	Exact		Approximate	
N	$P_1(N)$	$P_2(N)$	$P_1(N)$	$P_2(N)$
1	0.169 E+00	0.428 E+00	0.191 E+00	0.477 E+00
5	0.811 E-01	0.203 E+00	0.853 E-01	0.213 E+00
10	0.582 E-01	0.146 E+00	0.603 E-01	0.151 E+00
15	0.478 E-01	0.120 E+00	0.492 E-01	0.123 E+00
20	0.416 E-01	0.104 E+00	0.426 E-01	0.107 E+00
25	0.373 E-01	0.932 E-01	0.381 E-01	0.954 E-01
101	0.188 E-01	0.469 E-01	0.190 E-01	0.474 E-01
191	0.137 E-01	0.342 E-01	0.138 E-01	0.345 E-01

Then:

$$\pi_N(\mathbf{n}) = G(C)^{-1} \prod_{r \in R} \frac{(N\nu_r)^{n_r}}{n_r!} \qquad \mathbf{n} \in S(C)$$

where

$$S(N) = \{\mathbf{n} : A\mathbf{n} \leq NC\}$$

and

$$G(N) = \sum_{\mathbf{n} \in S(C)} \prod_{r \in R} \frac{(N\nu_r)^{n_r}}{n_r!}$$

Let $\mathbb{P}_r$ denote the blocking probability of calls along a given route r. Given the state spaces $S(N) = \{\mathbf{m} \in \mathbb{N}^R \, / \, \forall j, \, \sum_\rho A_{j\rho} m_\rho \leq NC_j\}$ and $\mathcal{K}(N) = \{\mathbf{k} \in \mathbb{N}^J : \forall j, \, K_j \leq NC_j\}$ for variables $\mathbf{m}$ and $\mathbf{K}$, respectively, define the sets $M_{jr} = \{\mathbf{m} \in S(N) : NC_j - A_{jr} < \sum_\rho A_{j\rho} m_\rho \leq NC_j\}$ and $\mathcal{K}_{jr} = \{\mathbf{k} \in \mathcal{K}(N) : NC_j - A_{jr} < K_j \leq NC_j\}$ corresponding equivalently to blocking states for route r on link j. We have

$$\mathbb{P}_r = p(\mathcal{M}_{1r} \cup \dots \cup \mathcal{M}_{Jr}) = q(\mathcal{K}_{1r} \cup \dots \cup \mathcal{K}_{Jr}).$$

Using Poincaré's formula for the probability of the union of events, we can write

$$\mathbb{P}_r = \sum_{l=1}^J (-1)^{l-1} \beta_l \tag{3.23}$$

with

$$\beta_l = \sum_{1 \leq j_1 < \dots < j_l \leq J} p(\mathcal{M}_{j_1,r} \cap \dots \cap \mathcal{M}_{j_l,r}) = \sum_{1 \leq j_1 < \dots < j_l \leq J} q(\mathcal{K}_{j_1,r} \cap \dots \cap \mathcal{K}_{j_l,r}) \tag{3.24}$$

depending on r and parameter N.

Now using the ideas of exponential twisting from the single resource problem and retaining the dominant terms in Poincaré's formula one can show the following main result for multi-rate loss networks under uniform conditions on links (all being lightly loaded, critical, or overloaded). We present the results without proof to indicate that they can be computed explicitly.

Proposition 3.9   *Consider the loss network with J links and R routes, in equilibrium with link capacities NC_j, route arrival rates $N\nu_r$ and a connection on route r holds $A_{j,r}$ bandwidth units on link j on route r. Let $A = \{A_{j,r}\}_{j=1,J;r=1,R}$ denote the link incidence matrix specifying the bandwidth requirements of all connections on the links with $A_{j,r} = 0$ if j is not on route r.*

Then we have the following estimates for the blocking probabilities on the routes.

Uniform Light Load:

Suppose that any $2 \times R$ sub-matrix of A has rank 2.

We then have

$$\mathbb{P}_r = \frac{1}{\sqrt{2\pi N}} \sum_{j, A_{jr} \neq 0} e^{-N.I_j} \frac{1}{\sqrt{\Gamma_j}} \frac{1 - e^{-y_j A_{jr}}}{1 - e^{-y_j}} \left[1 + O\left(\frac{1}{\sqrt{N}}\right)\right] \tag{3.25}$$

for large N, with

$$I_j = \sum_\rho \nu_\rho (1 - e^{-A_{j\rho} y_j}) - C_j y_j$$

and where

$$\Gamma_j = \sum_\rho \nu_\rho A_{j\rho}^2 e^{-A_{j\rho} y_j} \tag{3.26}$$

and $y_j < 0$ is the solution to:

$$\sum_\rho \nu_\rho A_{j\rho} e^{-A_{j\rho} y_j} = C_j. \tag{3.27}$$

Critical Load:

Assume that,

- *$\sum_r A_{jr} \nu_r = C_j$ for all $j = 1, 2, \ldots, J$*
- *and any $2 \times R$ sub-matrix of A has rank 2.*

We then have

$$\mathbb{P}_r = \frac{1}{\mathcal{H}^0 \sqrt{2\pi N}} \sum_{j, A_{jr} \neq 0} \frac{A_{jr}}{\sqrt{\Gamma_j}} \left[1 + O\left(\frac{1}{\sqrt{N}}\right)\right] \tag{3.28}$$

for large N, with $\Gamma_j = \sum_\rho \nu_\rho A_{j\rho}^2$ and

$$\mathcal{H}^0 = \mathbb{P}(W_1 \leq 0, \ldots, W_J \leq 0) \tag{3.29}$$

where $\mathbf{W}$ is the Gaussian vector in $\mathbb{R}^J$ with mean $\mathbf{0}$ and covariance matrix Γ.

Heavy Load:

Assume that the solution $\mathbf{y} \in \mathbb{R}^J$ *of system of equations*

$$\sum_\rho v_\rho A_{j\rho} e^{-A_{j\rho} y_j} = C_j. \tag{3.30}$$

has positive components for all $j = 1, 2, \ldots, J.$ *We then have*

$$\lim_{N \to +\infty} \mathbb{P}_r = 1 - \prod_{j, A_{jr} \neq 0} e^{-y_j A_{jr}}. \tag{3.31}$$

Remark 3.10   The critical and light load cases can be combined to obtain so-called uniform estimates. However, we can only obtain estimates under uniform loadings on all links. The general case with mixed loadings is very difficult to handle because we need to estimate the dominant term in Poincare's formula for computing the probability over the intersection of events k at a time where $k = 1, 2, \ldots, J.$ The rank conditions on the link incidence matrix $\mathbf{A}$ are related to the existence of solutions to the measure change problem that provides control on the order of the error terms. The details can be found in the paper of Simonian *et al* listed at the end of the chapter.

3.7   LOSS NETWORK MODELS IN WIRELESS COMMUNICATION NETWORKS

Loss network models also occur in the context of matching problems in multi-hop wireless networks. They are related to link activation. We briefly see how this occurs.

Consider a wireless network that is represented by a graph (V, E) where V denotes the set of vertices (or nodes) and E the set of edges that connect the vertices. Because wireless nodes share the same transmission medium they can interfere with each other if they simultaneously transmit. This causes the transmitted signal to be distorted and hence packets are lost. Thus, one of the challenges in wireless networks is to construct a so-called Medium Access protocol (MAC) that gives rules for which set of links in the graph can be simultaneously activated such that reliable transmission can be carried over that link. One simple protocol employed in practice (for example the one hop RTS/CTS scheme) is based on only allowing nodes at a certain distance (hop in terms of the graph structure) to transmit simultaneously— for example, only allowing nodes at a distance of at least k hops to transmit. This leads to the independent set problem in graphs– determining which edges (out of $|E|$) can be activated simultaneously. Let S denote the collection of all independent sets of (V, E) based on the $k \geq 2$ hop criterion assuming all links (edges between adjacent nodes at a 1 hop distance) are identical. Let $X_i \in \{0, 1\}$ denote the binary variable as to whether a link i is activated or not. Suppose nodes transmit packets on link i at a Poisson rate λ_i and they transmit for a duration that is random of length σ_i and $\rho_i = \lambda_i \mathbb{E}[\sigma_i]$. Assuming the durations are all independent, then let $X(t)$ denote the vector of links with $X_i(t) = 1$ if link i is transmitting at time t. Then it is easy to see that the equilibrium distribution of $X(t)$ is given by:

$$\pi(X) = \frac{1}{G} \prod_{i=1}^{|E|} \rho_i^{X_i}, \quad X \in S \tag{3.32}$$

where $G = \sum_{X \in S} \prod_{i=1}^{|E|} \rho_i^{X_i}$.

This is a simple loss network model of links with capacity 1 and requests requiring 1 unit of capacity. It is interesting to examine the behavior of the equilibrium distribution. Suppose the nodes have identical behavior, i.e., $\rho_i = \rho$. In that case, $\pi(X) = \frac{1}{G} \rho^{\sum_{X \in S} X_i}$, and thus this probability is maximized when $\sum_{X \in S} X_i$ is largest, i.e., X_i belongs to the largest independent set– called the maximal matching. Since this corresponds to the largest throughput assuming the load is ρ. Thus, we see that in the homogeneous case, the protocol model leads to the situation where the network is in the state corresponding to the maximal independent set or matching with highest probability. One implication of this result is that nodes at the *center* of a network are less likely to be in many independent sets and thus might suffer from unfairness of access. Thus, MAC scheduling algorithms that try to find the maximal matching for a given set of channel conditions (links that can be simultaneously ON) must also take into account the inherent unfairness arising from the geometry of the network. Much of the current research in so-called multi-hop ad hoc wireless networks focusses on trying to develop efficient distributed MAC schemes that solves the maximal matching problem given the random and fluctuating channel conditions while trying to provide fair access to all nodes.

3.8   SOME PROPERTIES OF LARGE MULTI-RATE SYSTEMS

We have seen that multi-rate Erlang loss systems with complete sharing (when all classes of arrival requests can access the available bandwidth without restrictions) result in stationary systems with product-form probabilities for the joint probability of number of calls in progress.

One of the nice consequences of the product-form structure is the *elasticity property* given below:

$$\frac{\partial \mathbb{P}_{B_m}}{\partial v_k} = \frac{\partial \mathbb{P}_{B_k}}{\partial v_m}, \quad 1 \le k, m \le R. \tag{3.33}$$

In other words, the sensitivity of the blocking of class m to changes in the class k rates is equal to the sensitivity of the class k blocking to the class m rates. However, as mentioned earlier there is no general monotonicity property for the blocking in multi-rate loss systems as was seen in the single rate classical Erlang loss formula i.e., increasing the rates of a given class does not necessarily increase the overall blocking in the system.

In large loss systems in the lightly loaded regime monotonicity does occur and this has important consequences in using the single-link approximations to develop Erlang fixed point methods as well as blocking sensitivity based routing algorithms. We now discuss these issues.

If we examine the formula for blocking in the uniform light load case, (3.25), we note that the blocking on route r is given by:

$$\mathbb{P}_{B_r} = \sum_{j, A_{j,r} \neq 0} B_{jr} \left(1 + O(\frac{1}{N}) \right) \tag{3.34}$$

where B_{jr} can be seen as the single link blocking formula for class r at link j that belongs to route r given by equation (3.20). We see that the blocking formula depends on the behavior of the rate function $I_j(C_j)$ in the first order approximation (in terms of N).

Lemma 3.11   *Let $I_j(C_j)$ be the rate function corresponding to the measure change on a link j with capacity C_j. Let v_r denote the arrival rate of a connection of type r on that link and assume that $\sum_{r:A_{j,r} \neq 0} v_r A_{jr} < C_j$ i.e., the link is lightly loaded. Then:*

$$\frac{\partial I_j(C_j)}{\partial v_r} < 0 \tag{3.35}$$

Proof. First note that by definition:

$$I_j(C_j) = \sum_{r:A_{jr} \neq 0} v_r (1 - e^{\tau_{C_j} A_{jr}}) + C\tau_{C_j}$$

where τ_{C_j} satisfies:

$$C_j = \sum_r e^{A_{jr} \tau_{C_j}} v_r A_{jr}$$

Noting that $\sum_{r:A_{jr} \neq 0} A_{jr} v_r < C_j$ we see that $\tau_{C_j} > 0$. Noting that from the definition:

$$\frac{\partial \tau_{C_j}}{\partial v_r} = \frac{-A_{jr} e^{A_{jr} \tau_{C_j}}}{\sum_r v_r A_{jr}^2 e^{A_{jr} \tau_{C_j}}}$$

Using this, the definition of $I_j(C_j)$ and the fact that $e^{A_{jr} \tau_{C_j}} > 1$ whenever $A_{jr} \neq 0$ we obtain:

$$\frac{\partial I_j(C_j)}{\partial v_r} = 1 - e^{A_{jr} \tau_{C_j}} < 0$$

$\square$

Using this result we can show the following monotonicity property for blocking in large loss networks that is not true in small networks.

Proposition 3.12

For a large loss network under uniform light loading conditions and N large enough, the following monotonicity property holds:

Let λ_j denote the vector of arrival rates ν_{jr} on link j.

Let λ'_j, λ_j be arrival rates on link $j \in J$, i.e., $\lambda_j = \sum_{r:A_{jr} \neq 0} \nu_r$ and B'_{jr}, B_{jr} be the corresponding blocking probabilities on that link. Assume that: $\lambda'_j \geq \lambda_j$, $j \in J$ (component wise) with at least one link i on which $\lambda'_i > \lambda_i$. Then $\mathbb{P}'_B > \mathbb{P}_B$ where $\mathbb{P}_B = col(\mathbb{P}_r)$, $r \in R$ is the vector of the blocking probabilities in the network.

Proof. The result follows from the previous lemma by noting that under the arrival rate conditions on link i we have $B'_{ir}(C_i, N) > B_{ir}(C_i, N)$ because the single link blocking is dominated by the term $e^{-N I_i(C_i)}$ and using the interpretation of the blocking along routes as the sum of the link blocking probabilities at each link on the route. □

Remark 3.13   The above conditions are trivially true if we increase the rate along any one route while keeping all other rates the same.

We end this section by showing that the elasticity property that holds for the exact formula is inherited by the asymptotic formulae.

Let B_{jr} denote the blocking at link j along route r. Clearly, from thr light load formula:

$$B_{jr} = \frac{e^{-N I_j(C_j)}(1 - e^{\tau C_j A_{jr}})}{\sqrt{N \sigma_j^2 (1 - e^{\tau C_j})}} \left(1 + O(\frac{1}{N})\right)$$

Once again if we dfferentiate B_{jr} with respect to ν_k where the flow k uses link j it can be shown (omitting the term in $O(\frac{1}{N})$):

$$\frac{\partial B_{jr}}{\partial N \nu_k} = -B_{jr}(1 - e^{\tau C_j A_{kj}})(1 + O(\frac{1}{N}))$$

$$= -\frac{e^{-N I_j(C_j)}(1 - e^{\tau C_j A_{jr}})(1 - e^{\tau C_j A_{kj}})}{\sqrt{N \sigma_j^2 (1 - e^{\tau C_j})}} \left(1 + O(\frac{1}{N})\right)$$

$$= \frac{\partial B_{jk}}{\partial N \nu_r} \left(1 + O(\frac{1}{N})\right)$$

establishing that the elasticity formula holds within a factor of $O(\frac{1}{N})$. Noting that the blocking along a route in light load is the sum of the link blocking probabilities it follows that the elasticity formula holds for large networks for the approximation formulae asymptotically.

The sensitivity of the blocking to rates can be used to develop load-based routing algorithms that are discussed in the monograph of Girard given in the reference list. However, as has been seen the *largeness* of systems in terms of their server rates to individual connection requirements can be used to our advantage from a computational perspective as we can obtain explicit formulae for blocking as well as their sensitivities with respect to traffic and capacity parameters.

CONCLUDING REMARKS

In this chapter we have seen that we can obtain very explicit results for loss system models that have a *product-form* structure resulting from the fact that the underlying stationary probability distribution is related to $M/G/\infty$ models restricted to a state-space that is a closed convex set. There exist many models where the state-space is not a closed convex set. In such a case we cannot exploit these techniques. Examples of such situations occur in loss systems with bandwidth or trunk reservation, i.e., where a certain part of the state space is accessible only by certain types of connections.

Loss models might appear very special however they are immensely useful as models for dimensioning real systems. In the next chapter we will see that there is a way of reducing queueing based models to loss models via the notion of *effective bandwidths*.

NOTES AND PROBING FURTHER

The history of loss systems goes back to the work of Erlang, Engset and Palm who were Scandinavian telecommunication engineers in the early 1900's. The results were for classical telephone networks where each call held only one circuit at a time on a link or switch. The need for the multi-rate extension arose with the advent of ISDN (Integrated services Digital Networks) when dial-up modems first came into use where a connection could request from a fixed set of bandwidths (3 initially) depending on whether it was a voice or data connection. The extension to the many classes case is natural in the networking context because the number of classes are defined on origin-destination pairs with fixed routes between them and link speeds could be different on different links. There are many variations of these models referred to as stochastic knapsack problems because of the random arrival and sojourn of calls or connections. Loss networks play an important role in optical networks where the circuit multiplexing paradigm is natural because of the nature of the transmission mechanism (it is difficult to change wavelengths along the way). More recently loss network models have found a role in the calculation of blocking in backbone of the Internet with MPLS (Multiprotocol Label Switching) where-by links are virtual and capacity along paths is reserved.

Finally, we have only discussed the case when the distribution has a product-form. In many contexts (including optical networks) the stationary distributions do not have product-form. In that case, under assumptions of Poisson arrivals and exponential holding times there are algorithmic approaches to solve for the steady state of the underlying Markov chain. In many cases, even though a product-form might not exist, if the systems are large, Kaufman-Roberts type recursions can be used in sub-regions of the state-space with some way to patch the boundary probabilities. Such methods do give accurate estimates for blocking probabilities in the case of loss systems with trunk reservation for example.

BOOK REFERENCES

Loss network models are discussed in many books. Some are purely interested in the structure of the distributions and reversibility issues, others use the blocking formulae for network dimensioning and optimization in the context of state-dependent least-loaded routing algorithms. These methods are extremely useful and many current networks are based on such routing algorithms because the modeling assumptions are fairly robust.

F. P. Kelly:*Reversibility and Stochastic Networks*, J. Wiley and Sons., Chichester, 1979.

This is an excellent classic reference for understanding general Markovian models of networks, reversibility, and insensitivity.

K. W. Ross: *Multiservice Loss Models for Broadband Telecommunication Networks (Telecommunication Networks and Computer Systems)*, Springer-Verlag, 1995.

This is a comprehensive and excellent reference on loss systems and networks. There is a detailed discussion of the stochastic knapsack problem and algorithmic issues.

A. Girard,*Routing and Dimensioning in Circuit-Switched Networks*, Addison-Wesley Longman Publishing Co., Inc., Boston, MA, 1990

This book presents a comprehensive view of optimization based on the loss network model for routing and dimensioning.

A. Kumar, D. Manjunath, and J. Kuri: *Communication Networking: An analytical approach*, Morgan-Kaufman (Elsevier), 2004.

This recent text is an useful source on circuit multiplexing and loss network models. A detailed presentation of aspects of loss networks such as traffic overflow characteristic etc. are given along with the Erlang fixed point approach.

JOURNAL ARTICLES

Loss networks continues to attract the attention of many researchers. There are a number of good survey articles on the topic two of which are listed at the end.

J. S. Kaufman, *Blocking in a shared resources environment*, IEEE Trans. Commun., COM-29, 10, 1981, pp.1474-1481.

J. W. Roberts, *A service system with heterogeneous user requirements: Application to multi-service telecommunications systems*, Perf. of Data Comm. Syst. and their Applications, G. Pujolle ed., pp.423-431, North-Holland, 1981.

Z.Dziong, J.W.Roberts (1987), *Congestion probabilities in a circuit switched integrated services network*, Performance Evaluation, Vol.7, N° 3.

The two papers by J. Kaufman and J. R. Roberts appeared almost simultaneously. Both of them contain the Kaufman-Roberts algorithm for single resource systems. The third paper develops the Kaufman-Roberts algorithm for the network case.

F.P. Kelly, *Blocking probabilities in large circuit-switched networks*, Adv.Appl.Prob. **18**, 1996, pp.473-505

In this paper Kelly introduced the idea of the Erlang fixed point and showed that it is asymptotically accurate. In addition to these results Kelly obtained the heavy-load limit given in Proposition 3.9.

G. Louth, M. Mitzenmacher and F.P. Kelly, *Computational complexity of loss networks*, Theoretical Computer Science, Vol. 125 (1), 1994, pp. 45-59.

This paper presents the difficulty in computing the blocking probability in loss models in terms of the computations that are required based on the structure of the blocking states. These results led to much activity to develop methods for large networks.

P. Nain, *Qualitative properties of the Erlang blocking model with heterogeneous user requirements* , Queueing Systems, Vol. 6 (1), 1990, pp. 189-206.

P. Gazdzicki, I. Lambadaris, R.R. Mazumdar, *Blocking probabilities for large multi-rate Erlang loss systems*, Adv.Appl.Prob. 25, 1993 pp. 997-1009

A. Simonian, F. Théberge, J. Roberts, and R. R. Mazumdar, *Asymptotic estimates for blocking probabilities in a large multi-rate loss network*, Advances in Applied Probability, Vol. 29, No. 3, 1997, pp. 806-829.

D.Mitra, J.Morrison, *Erlang capacity and uniform approximations for shared unbuffered resources*, IEEE/ACM Transactions on Networking, vol.2, N.6, 1994, pp.558-570

The first two references should be consulted for the large loss systems and networks analysis presented in this chapter based on a local limit large deviations approach. The third reference derives the same results as in Gazdicki et. al by a complex analytic approach using saddle-point approximations based on computing the moment generating function of the stationary distribution.

G. L. Choudhury, D. L. Lucantoni, and W. Whitt, *Numerical Transform Inversion to Analyze Teletraffic Models.*, The Fundamental Role of Teletraffic in the Evolution of Telecommunications Networks, Proceedings of the 14th International Teletraffic Congress, J. Labetoulle and J. W. Roberts (eds.), Elsevier, Amsterdam, vol. 1b, 1994, 1043-1052.

This paper presents another approach via the use of inversion techniques of the moment generating function of stationary distributions. The third author has pursued this issue using many different inversion techniques for many types of models.

S. P. Chung and K. W. Ross, *Reduced load approximation for multi-rate loss networks*, IEEE Trans. on Communications, COM-41 (8), 1993, pp. 1222-1231.

F. Theberge and R.R. Mazumdar(1996), *A new reduced load heuristic for computing blocking in large multirate loss networks*, Proc. of the IEE- Communications, Vol 143 (4), 1996, pp. 206-211.

These two papers address the extension of the Erlang fixed point via the use of the single-link multi-rate loss models in the large network case. The first paper uses for a single link the multi-rate Erlang loss model that is approximated by so-called Monte-Carlo summations. They show that in this case a unique fixed point may not exist. The second paper shows that the large loss model yields accurate results for even finite N and asymptotically exact results. Moreover, it proves convergence to a unique solution by using the iteration technique as in the Erlang fixed point.

F. P. Kelly, *Loss networks*, Annals of Applied Probability, 1 (1991), 319-378

This is an excellent survey on loss networks where many issues such as the dynamic behavior as well as the critically loaded regime are discussed along with the Erlang fixed point ideas.

S. Zachary and I. Zeidins, *Loss Networks*, May 2009, available from `http://arxiv.org/abs/0903.0640v1`

This is a more up to date survey of the state-of-the art in loss networks. A different approach to obtain blocking in large networks based on he Kaufman-Roberts recursion is presented which offers a different viewpoint of large loss networks by concentration on most likely states.

CHAPTER 4

Stochastic Networks and Insensitivity

4.1 INTRODUCTION

Loss networks that we discussed in the previous chapter are very special types of stochastic networks where the stationary joint distribution has a *product form* that is completely characterized by the (Poisson) arrival rates of traffic on different routes and the mean of the holding times of the calls or sessions. Essentially this *insensitive* property comes from the fact that the system is a set of infinite server queues restricted to a closed subspace and so it inherits both the *product form* structure as well as insensitivity.

In this chapter we will study these issues further to see whether there are more general classes of stochastic networks for which these properties hold. It turns out that this is very closely related to a notion of partial balance of Markov chains that we will encounter. Furthermore, recent work has also shown that there is a very nice connection between insensitivity of stochastic networks and bandwidth sharing strategies or how the server capacity is used. This chapter is devoted to a study of these issues.

4.2 MARKOVIAN NETWORKS

Markovian networks refer to networks of queues with external Poisson arrivals and independent exponentially service times at the various queues. Such network models are more of mathematical interest rather than modeling reality since in real networks an arriving job, packet, etc., carries its job size with it as it progresses through the network and thus the service times at the various queues are not independent. Let us begin by considering the simplest case first.

Consider a set of tandem queues as shown in Figure 4.1. At each queue $i, i = 1, 2, \ldots, N$ except there are two types of arrivals. The first are arrivals from queue $i - 1$ and external arrivals that are Poisson with rate λ_i. At each queue i the service times of the packets are i.i.d. exponentially distributed random variables with mean $\frac{1}{\mu_i}$. All packets leave from the last queue N.

Let $\mathbf{n} = (n_1, n_2, \ldots, n_N)$ denote the vector of the number of packets in each queue in equilibrium. Define $\rho_i = \frac{\sum_{j=1}^{i} \lambda_j}{\mu_j}$. Under the condition $\max_{1 \le i \le N} \rho_i < 1$, a stationary distribution exists

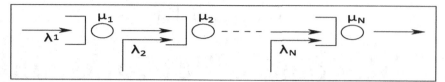

Figure 4.1: Tandem queue.

and is given by:

$$\Pi(\mathbf{n}) = \prod_{i=1}^{N}(1 - \rho_i)\rho_i^{n_i} \tag{4.1}$$

This is very easy to see. By repeated application of Burke's theorem and the fact that the sum of independent Poisson processes is Poisson with a rate equal to the sum of the rates, every output process is Poisson with the same rate as the input. Then because of the assumption of i.i.d. exponential service times, each queue is a $M/M/1$ queue with traffic intensity ρ_i independent of the others in equilibrium and hence the result follows. This model easily generalizes to a tandem model where some packets can exit from queue i with probability p_i or go to the next queue with probability $1 - p_i$ in an independent manner. All that changes is that the Poisson rates that enter queue i denoted by Λ_i change as: $\Lambda_i = (1 - p_{i-1})\Lambda_{i-1} + \lambda_i$ and the same result (4.1) holds by defining $\rho_i = \frac{\Lambda_i}{\mu_i}$.

Now suppose we complicate this model a bit by allowing the output of a queue i to be fed back to j with $j \leq i$. All of a sudden things get more interesting because the input processes to queues are no more Poisson even if the service times are exponential. Let us see this by considering a simple $M/M/1$ queue with feedback. Here the external arrivals are Poisson with rate λ, the services are i.i.d. exponential with mean $\frac{1}{\mu}$. Upon service completion, with probability p a job is fed back and with probability $(1 - p)$ it leaves the system. Let us study this model in detail. It is illustrated in Figure 4.2.

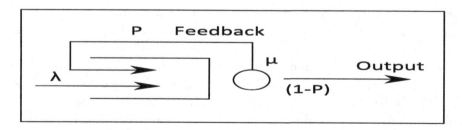

Figure 4.2: Queue with feedback.

Let Λ denote the average rate of arrivals (external and feedback) into the queue. From stability the output rate is Λ and hence the rate at which packets are fed back is Λp.

Therefore, $\Lambda = \lambda + p\Lambda$ or $\Lambda = \frac{\lambda}{1-p}$. Clearly, for stability, $\Lambda < \mu$ or $\lambda < (1-p)\mu$. Now by solving the balance equations for the Markov chain it is easy to see that the stationary distribution, denoted by $\pi(n)$ is given by:

$$\pi(n) = \left(\frac{\lambda}{(1-p)\mu}\right)^n \pi(0) = \left(\frac{\Lambda}{\mu}\right)^n \pi(0) \qquad (4.2)$$

where $\pi(0) = 1 - \frac{\lambda}{(1-p)\mu}$.

This is just the form of the stationary distribution for a $M/M/1$ queue with input rate $\Lambda = \frac{\lambda}{1-p}$. It is important to note that the total input process to the queue is not Poisson, however the output process leaving the system is still Poisson with rate λ the same as the external input rate! Let us show this.

Let $\{D_n^0\}$ denote the external departures and $\{D_n^1\}$ denote the feedback point process. Let $\rho = \frac{\lambda}{(1-p)\mu} = \mathbb{P}(Q_0 > 0)$ and from the fact that external arrival and departure distributions coincide, from PASTA, $\mathbb{P}_D(Q_D > 0) = \mathbb{P}(Q > 0) = \rho$.

Following the same arguments as in the proof of the departures in a $M/M/1$ queue (cf. Theorem 2.24) we obtain:

$$
\begin{aligned}
\mathbb{E}[e^{-h(D_{n+1}^0 - D_n^0)}] &= \mathbb{E}[e^{-h(D_{n+1}^0 - D_n^0)} | Q_{D_n} = 0]\mathbb{P}(Q_{D_n} = 0) + \\
&\quad \mathbb{E}[e^{-h(D_{n+1}^0 - D_n^0)} | Q_{D_n} > 0]\mathbb{P}(Q_{D_n} > 0) \\
&= \frac{\mu(1-p)}{h + (1-p)\mu}\rho + \mathbb{E}[e^{-h(D_{n+1}^0 - T_{n+1} - D_n^0 - T_{n+1})} | Q_{D_n} = 0](1-\rho)
\end{aligned}
$$

where we note by the exponential assumption of the services, when there are packets in the queue the output rate is just an independent sampling with probability $(1-p)$ of exponential r.v's that are exponential with rate $\mu(1-p)$. Also when there are no packets in the queue the arrivals correspond to external arrivals that are Poisson, hence $T_{n+1} - D_n^0$ is distributed as exponential λ. Moreover, $D_{n+1}^0 - T_{n+1}$ and $T_{n+1} - D_n^0$ are independent.

Therefore, substituting all the values we obtain:

$$
\begin{aligned}
\mathbb{E}[e^{-h(D_{n+1}^0 - D_n^0)}] &= \left(\frac{\mu(1-p)}{h + \mu(1-p)}\right)\frac{\lambda}{\mu(1-p)} + \\
&\quad \frac{\lambda}{\lambda + h}\left(\frac{\mu(1-p)}{h + \mu(1-p)}\right)(1 - \frac{\lambda}{(1-p)\mu}) \\
&= \frac{\lambda}{h + \mu(1-p)}\left(1 + \frac{\mu(1-p) - \lambda}{h + \lambda}\right) = \frac{\lambda}{h + \lambda}
\end{aligned}
$$

Noting that the r.h.s. corresponds to the Laplace transform of an exponential distribution with parameter λ, it implies that the external departures are points of a Poisson process with intensity λ.

If we repeat the same argument for the feedback arrivals we see that they are not Poisson. Indeed, we can show that:

$$\mathbb{E}[e^{-h(D_{n+1}^1 - D_n^1)}] = \frac{c_1}{h+\lambda} + \frac{c_2}{h+\mu p}$$

where c_1 and c_2 are constants showing that the inter-arrival times of the feedback process is hyper-exponential and not Poisson. Hence, the total arrival process to the queue is not Poisson.

Thus, although the arrivals to the queue are not Poisson the queue behaves as a $M/M/1$ queue with Poisson arrival rate λ and service rate $\mu(1-p)$. If we look at the Markov process $\{Q_t\}$ then in light of equation (4.2) we see that the process satisfies the reversibility condition w.r.t the external arrival and departure rates λ and $\mu(1-p)$ respectively.

It turns out that this property holds more generally for a class of networks that satisfy a property called *quasi-reversibility* due to Kelly. The idea of quasi-reversibility is closely related to reversibility of stationary Markov chains discussed in the Appendix. We first state an important result due to Kelly.

Lemma 4.1   *Let $\{X_t, t \in \mathbb{R}\}$ be a stationary CTMC (continuous-time Markov chain) on $E \times E$ with generator $Q = \{q_{ij}\}$. If we can find a set of non-negative number $\tilde{q}_{ij}$, $i, j \in E$ and non-negative numbers π_j, $j \in E$ with $\sum_{j \in E} \pi_j = 1$ such that:*

$$\sum_{j \neq i} \tilde{q}_{ij} = \sum_{j \neq i} q_{ij} \ \forall i \in E \tag{4.3}$$

$$\pi_j q_{ji} = \pi_i \tilde{q}_{ij} \ \forall i, j \in E \tag{4.4}$$

then $\tilde{Q} = \{\tilde{q}_{ij}\}$ is the generator for the reversed process and π is an invariant (stationary) distribution for both the forward and reversed processes.

Proof. Let π and Q satisfy the conditions above, then for each $i \in E$ we have:

$$\sum_{j \neq i} \pi_j q_{ji} = \sum_{j \neq i} \pi_i \tilde{q}_{ij} \ from \ (4.4)$$

$$= \pi_i \sum_{j \neq i} q_{ij} \ from \ (4.3)$$

$$= -\pi_i q_{ii}$$

since $\sum_{j \in E} q_{ij} = 0$ or $q_{ii} = -\sum_{j \neq i} q_{ij}$. Therefore, we have $\pi Q = 0$ or π is the stationary distribution of X_t.

It remains to show that the reversed process is Markov with the generator matrix $\tilde{Q}$. This readily follows from the results in the Appendix since:

$$\tilde{q}_{ji} = \frac{\pi_i q_{ij}}{\pi_j} \tag{4.5}$$

and $\pi \tilde{Q} = 0$ also showing that π is the stationary distribution of the reversed process.   □

We now state and prove results for an important class of Markovian stochastic networks called single-class Jackson networks. By single class it is meant that all packets or customers at a given node are treated in a similar manner in that the service times are i.i.d. exponential. Packets are not differentiated by other attributes as for example in a multi-class network where one type of customer is given preference over the other, etc. This is depicted in Figure 4.3.

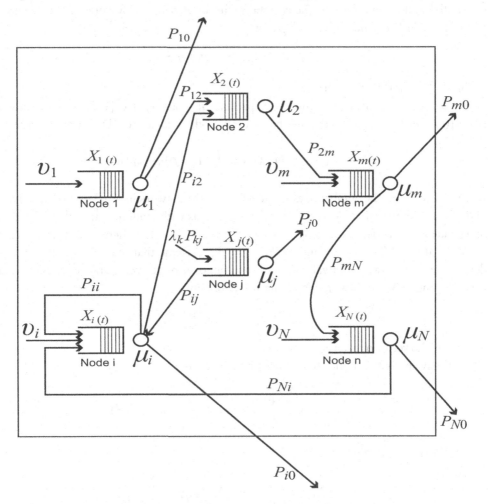

Figure 4.3: Open Markovian network model.

Consider a stochastic network formed of N nodes. External arrivals to a node i, $i \in \{1, 2, \ldots, N\}$ are Poisson with rate v_i. Service times of packets at a node i are i.i.d. exponential with rate μ_i. After service at node i a packet is routed independently to node j, $j \in \{1, 2, \ldots, N\}$ with probability p_{ij} and leaves the network with probability $p_{i0} = 1 - \sum_{j=1}^{N} p_{ij}$. Note if $p_{i0} > 0$

for at least one $i \in \{1, 2, \cdots, N\}$ we say the network is *open*. If $p_{i0} = 0$ $\forall i$ then the network is said to be *closed*.

Let P denote the routing matrix with p_{ij} denoting the probability of a packet being routed to node j after service a node i. Then the assumption that the network is open implies that for every i, $\sum_{j=1}^{N} p_{ij} \leq 1$ and $\sum_{j=1}^{N} p_{ij} < 1$ for at least one i. This assumption implies that P is sub-stochastic[1] and is equivalent to the spectral radius of P denoted by $sp(P)$ (and therefore of P^T) being less than 1. Furthermore, noting that:

$$(I - P)^{-1} = I + P + P^2 + \cdots$$

we have $\|(I - P)^{-1}\| < \infty$. a

Let $x_i(t)$ denote the number of packets at node i, $i \in \{1, 2, \cdots, N\}$ and let $X_t = col\,(x_1(t), x_2(t), \cdots, x_N(t))$ denote the vector of node occupancies. Then we can write:

$$X_t = X_0 + A(t) - \int_0^t (I - P^T(s))dD(s \qquad (4.6)$$

where $A(t)$ denotes the number of external arrivals to the network in $(0, t]$, $D(t)$ denotes the number of departures from each node, and $P^T(t)$ denotes matrix with random entries that take values $\{0, 1\}$, independent of the departure times with $P_{ij}^T(t) = 1 = P_{ji}(t)$, if there is a departure from node j to node i at t and 0 otherwise and $\mathbb{E}[P_{ij}(t)] = p_{ij}$. Assume that the network is stable, and let λ_i denote the average rate of arrivals at node i, and hence also the average rate of departures from that node. Noting $\mathbb{E}[X_t] = \mathbb{E}[X_0]$, in stationarity we obtain:

$$0 = v + (I - P^T)\lambda \qquad (4.7)$$

where v is the vector of external arrival rates.

Equivalently the traffic equation can be written as:

$$\lambda_i = v_i + \sum_{j=1}^{N} \lambda_j p_{ji}, \quad i \in \{1, 2, \cdots, N\} \qquad (4.8)$$

Moreover, note that:

$$\sum_{i=1}^{N} v_i = \sum_{j=1}^{N} \lambda_j (1 - \sum_{i=1}^{N} p_{ji}) = \sum_{j=1}^{N} \lambda_j p_{j0} \qquad (4.9)$$

Hence, noting $sp(P) < 1$, for open networks there always exists a solution to the traffic equation (4.7) given by $\lambda = (I - P^T)^{-1}v$.

Proposition 4.2 Jackson Networks

[1] Recall a matrix with non-negative entries is said to be stochastic if all row sums are 1.

Consider an open network with N FIFO nodes. Suppose the external arrival rate be Poisson with rate v_i at node i and let $P = \{p_{ij}\}$ denote the routing matrix whose spectral radius is < 1. Suppose the service times at node i are i.i.d exponential with rate μ_i, $i \in \{1, 2, \cdots, N\}$. Under the assumption that $\lambda_i < \mu_i$ where λ_i solves the traffic equation, the Markov process $\{X_t, t \in \mathbb{R}\}$ is positive recurrent and the stationary distribution denoted by $\pi(x)$ is given by:

$$\pi(x) = \prod_{i=1}^{N}(1 - \rho_i)\rho^{x_i} \tag{4.10}$$

where $\rho_i = \frac{\lambda_i}{\mu_i}$

Proof. To show this result we will invoke the results of Lemma 4.1. To simplify the proof let us assume that $p_{i0} > 0$, the probability that a packet leaves the system after service at node i is non-zero. Let us first see the structure of the generator matrix of the Markov chain. Let e_i denote the unit vector with a "1" in the ith. place and 0 elsewhere. Let $q(x, y)$ denote the rate from state x to state y. Let $\mathbf{n} = col(n_1, n_2, \cdots, n_N)$. Then:

$$q(\mathbf{n}, \mathbf{n} + e_i) = v_i \tag{4.11}$$
$$q(\mathbf{n}, \mathbf{n} - e_i) = \mu_i p_{i0} \mathbb{1}_{[n_i > 0]} \tag{4.12}$$
$$q(\mathbf{n}, n - e_i + e_j) = \mu_i p_{ij} \mathbb{1}_{[n_i > 0]} \tag{4.13}$$

Let $\tilde{v}_i$ denote the average rate of departures from node i outside the system. If the queue is stable then:

$$\tilde{v}_i = \lambda_i p_{i0} \tag{4.14}$$

If this does not hold true then there will be accumulation at the node and the system cannot be stable. Let us now invoke the idea of time reversibility. If the system is stable, then in equilibrium, the rates of arrival and departure in the forward and reverse directions must be the same. However, the routing probabilities must change since the rate at which packets go from i to j in forward time must be the same as the rate at which packets go from j to i in reverse time. Let $\tilde{p}_{ij}$ denote the routing matrix. Then

$$\lambda_j \tilde{p}_{ji} = \lambda_i p_{ij} \implies$$
$$\tilde{p}_{ji} = \frac{\lambda_i p_{ij}}{\lambda_j}$$

Having identified the new routing matrix for the reverse process it can be readily checked that it is also sub-stochastic. Indeed, noting that $\lambda_j = v_j + \sum_{i=1}^{N} p_{ij}\lambda_i$:

$$\sum_{i=1}^{N} \tilde{p}_{ji} = \frac{1}{\lambda_j} \sum_{i=1}^{N} p_{ij}\lambda_i = \frac{\sum_{i=1}^{N} p_{ij}\lambda_i}{v_j + \sum_{j=1}^{N} p_{ij}\lambda_i} < 1$$

It now remains to check that the conditions of Lemma 4.1 hold. For this we can check that:

$$\sum_{\mathbf{m}\neq\mathbf{n}} \tilde{q}(\mathbf{n}, \mathbf{m}) \;=\; \sum_{\mathbf{m}\neq\mathbf{n}} q(\mathbf{n}, \mathbf{m}) \quad \forall \mathbf{n}, \mathbf{m} \in \mathbb{Z}_+^N \tag{4.15}$$

$$\pi(\mathbf{n})q(\mathbf{n}, \mathbf{m}) \;=\; \pi(\mathbf{m})\tilde{q}(\mathbf{m}, \mathbf{n}) \quad \forall \mathbf{n}, \mathbf{m} \in \mathbb{Z}_+^N \tag{4.16}$$

Let us first verify that (4.3) is satisfied. By the definition of $\tilde{q}(\mathbf{n}, \mathbf{m})$ we have, by appropriate definition of the $\tilde{q}(x, y)$ as in (4.11), (4.12), and (4.13) with $\tilde{v}_i, \tilde{\mu}_i$, and $\tilde{p}_{ij}$ in place of v_i, μ_i, and p_{ij} respectively.

$$\begin{aligned}
\sum_{\mathbf{m}\neq\mathbf{n}} \tilde{q}(\mathbf{n}, \mathbf{m}) &= \sum_{i,j\in\{1,2,\cdots,N\}} \left(\tilde{q}(\mathbf{n}, \mathbf{n} - e_i) + \tilde{q}(\mathbf{n}, \mathbf{n} - e_i + e_j) + \tilde{q}(\mathbf{n} + e_i) \right) \\
&= \sum_{i=1}^{N} \left(\tilde{v}_i + \tilde{\mu}_i \mathbb{1}_{[n_i>0]} \right) \left(\tilde{p}_{i0} + \sum_{j=1}^{N} \tilde{p}_{ij} \right) \\
&= \sum_{i=1}^{N} \left(\lambda_i p_{i0} + \mu_i \mathbb{1}_{[n_i>0]} \right) \\
&= \sum_{i=1}^{N} \left(v_i + \mu_i \mathbb{1}_{[n_i>0]} \right) = \sum_{\mathbf{m}\neq\mathbf{n}} q(\mathbf{n}, \mathbf{m})
\end{aligned}$$

We now need to show that condition (4.4) is verified. Let us do it for $\mathbf{n}$ and $\mathbf{m} = \mathbf{n} + e_i$ using the form given by $\pi(n)$ in (4.10).

By definition $\tilde{q}(\mathbf{n}, \mathbf{n} + e_i) = \tilde{v}_i = \lambda_i p_{i0}$ and $q(\mathbf{n} + e_i, \mathbf{n}) = \mu_i p_{i0} \mathbb{1}_{[\mathbf{n}+e_i>0]} = \mu_i p_{i0}$. Substituting into (4.4) we obtain:

$$\begin{aligned}
\pi(\mathbf{n})\tilde{q}(\mathbf{n}, \mathbf{n} + e_i) &= \prod_{j=1}^{N} (1 - \rho_j)\rho^{n_j} \lambda_i p_{i0} \\
&= \prod_{j\neq i} (1 - \rho_j)\rho_j^{n_j}(1 - \rho_i)\rho_i^{n_i+1} \mu_i p_{i0} \\
&= \pi(n + e_i)\mu_i p_{i0} = \pi(n + e_i)q(n + e_i, \mathbf{n})
\end{aligned}$$

In a similar way we can show that $\pi(\mathbf{n})\tilde{q}(\mathbf{n}, \mathbf{n} - e_i + e_j) = \pi(n + e_j - e_i)q(n + e_j - e_i, \mathbf{n})$, thus showing that $\pi(\mathbf{n})$ is the stationary distribution by Lemma 4.1. $\square$

Remark 4.3   As in the simple case of a $M/M/1$ queue with feedback, it can be shown that the internal flows within the network are not Poisson. However, the external departure process from a given node i is Poisson with rate $\lambda_i p_{i0}$. Moreover, from the product-form of the stationary distribution, in equilibrium the number or packets at each node is mutually independent of the other nodes (as random variables and not processes!) each behaving as a $M/M/1$ queue with arrival rate λ_i and service rate μ_i. Hence, the sojourn time at a given node, denoted by T_i is given by:

$$\mathbb{E}[T_i] = \frac{1}{\mu_i - \lambda_i}$$

Note here the mean is with respect to the stationary distribution even though the process of arrivals to a node is not Poisson, i.e., the PASTA property holds for all flows within the network.

Noting that $((I - P)^{-1}$ is well defined, it means that the probability that a packet entering the network has a finite sojourn within the network since $\sum_{n=1}^{\infty}(P^n)_{ij}$ is the expected number of visits of a packet leaving node i to node j which is finite. From this we can compute the total expected sojourn time of a typical packet in the network, denoted by W from Little's formula, noting $\lambda = \sum_{i=1}^{N} \lambda_i = \sum_{i=1}^{N} ((I - P^T)^{-1} v)_i$, as

$$
\begin{aligned}
\mathbb{E}[W] &= \frac{1}{\lambda} \sum_{i=1}^{N} \mathbb{E}[X_i] = \frac{1}{\lambda} \sum_{i=1}^{N} \frac{\lambda_i}{\mu_i - \lambda_i} \\
&= \sum_{i=1}^{N} \frac{1}{\mu_i - \lambda_i} \sum_{j=1}^{N} \left((I - P)^{-1} - I \right)_{ji}
\end{aligned}
$$

In the above treatment we assumed that the network was *open*. We now discuss the question of closed Jackson networks. These are sometimes referred to as Gordon-Newell networks. Here the main difference is that the routing matrix is stochastic and there are no external arrivals because if external arrivals cannot exit the network the state of the system would drift to infinity and there would be no stationary distribution.

The traffic equation of a closed network is given by:

$$
\lambda_i = \sum_{j=1}^{N} \lambda_j p_{ji}, \quad i = 1, 2, \cdots, N \tag{4.17}
$$

Note the solution of the traffic equation can be obtained up to a constant.

We now state the result. The proof follows by checking that the conditions of Kelly's lemma hold and the existence follows from the fact that it is a finite-state Markov chain.

Proposition 4.4 Closed Jackson Network

Consider a closed Jackson network with N nodes and M packets in the system and service times at node i being i.i.d. exponential with rate μ_i. Then the Markov process associated with the numbers of packets at each queue is positive recurrent and the stationary distribution for $x = (x_1, x_2, \cdots, x_N)$ denoted by $\pi(x)$ is given by:

$$
\pi(x) = \frac{1}{G_M} \prod_{i=1}^{N} \rho_i^{x_i}, \quad x \in \mathbb{Z}_{=}^{N} : \sum_{i=1}^{N} x_i = M \tag{4.18}
$$

where G_M is a normalization constant and $\rho_i = \frac{\lambda_i}{\mu_i}$.

Note that unlike the open Jackson network case the queue occupancy processes are not independent.

One of the nice properties of closed networks is that knowledge of the normalizing constant G_M is useful for computing many quantities of interest. The stationary distribution can be written in terms of the normalization constant as:

$$\pi(X_i = m) = \frac{G_{M-m} - \rho_i G_{M-m-1}}{G_M} \rho_i^m$$

The tail distribution of a node, say i, is given by:

$$\mathbb{P}(X_i \geq m) = \rho_i^m \frac{G_{M-m}}{G_M}, \quad m = 1, 2, \cdots, M \tag{4.19}$$

Using this one can readily calculate various performance quantities such as the average throughput at a node. Let $\gamma_i(M)$ denote the throughput of node i

$$
\begin{aligned}
\gamma_i(M) &= \mathbb{P}(x_i > 0)\mu_i \\
&= \mu_i \rho_i \frac{G_{M-1}}{G_M} \quad \textit{from 4.19 with } m = 1 \\
&= \frac{G_{M-1}}{G_M} \lambda_i
\end{aligned}
$$

Since the closed queueing network has a finite number of customers, the arrivals are not Poisson.

Proposition 4.5 Arrival Theorem for Closed Jackson Networks

Consider a closed Jackson network with M number of customers with arrival rate λ_i and service rate μ_i. Then the stationary distribution seen by an arrival to queue i, $i = 1, 2, \cdots, N$ is the stationary distribution of a Jackson network with $M - 1$ customers, i.e.,

$$\pi_A(\mathbf{x}) = \frac{1}{G_{M-1}} \prod_{i=1}^{N} \rho^{x_i}, \quad s.t. \sum_{i=1}^{N} x_i = M - 1 \tag{4.20}$$

Proof. Consider a closed network N nodes and M customers. When a customer has received all the service required in station (queue) i, it joins station j with probability p_{ij}. The service times and the routing random variables are independent. The service time at station i is exponential with mean μ_i^{-1}. In each station there is one server working at unit speed, and a waiting room of capacity of larger than M. Therefore, defining

$$X(t) = \big(X_1(t), \ldots X_N(t)\big); \quad \sum_{i=1}^{N} X_i(t) = M$$

where $X_i(t)$ is the number of customers in station i (waiting or being served) at time t. The infinitesmal characteristics of the Markov chain is given by

$$q_{\mathbf{n},\mathbf{n}-e_i+e_j} = \mu_i \, p_{ij} \mathbf{1}_{[n_i>0]}$$

If the stochastic matrix $P = \{p_{ij}\}$ is irreducible and all the μ_i's are strictly positive, the chain $\{X(t)\}$ is irreducible, and since the state space is finite, it is ergodic.

Let N_{ij} be the counting process counting the transfers from station i to station j. It admits the $\mathcal{F}_t^X$-intensity

$$\lambda_{ij}(t) = \mu_i \, p_{ij} \, \mathbf{1}_{[X_i(t)>0]}$$

and by the Papangelou's formula

$$\lambda_{ij} \mathbb{P}_{N_{ij}}^0 [X(0-) = \mathbf{n}] = \mathbb{E}\left[\mu_i \, p_{ij} \mathbf{1}_{[X_i(t)>0]} \mathbf{1}_{[X(0-)=\mathbf{n}]}\right]$$

where $\lambda_{ij} = \mu_i \, p_{ij} \, P(X_i(0) > 0)$ is the average intensity of N_{ij}. Therefore,

$$\begin{aligned}
\mathbb{P}_{N_{ij}}^0 [X(0-) = \mathbf{n}] &= \frac{1}{C} \Big(\prod_{l=1, l\neq i}^{N} \rho_l^{n_l} \Big) \rho_i^{n_i-1} \quad if \quad n_i > 0 \\
&= 0 \quad \text{otherwise}
\end{aligned}$$

where C is the constant of normalization given by

$$C = \sum_{n_1+...+n_N=M, n_i>0} \Big(\prod_{l=1, l\neq i}^{N} \rho_l^{n_l} \Big) \rho_i^{n_i-1}$$

But the right-hand side of the above equality, after the change of summation variable $n_i \to n_i - 1$ is just G_{M-1}. Therefore,

$$\mathbb{P}_{N_{ij}}^0 [X(0-) = \mathbf{n}] = \pi_A(\mathbf{n}) = \frac{1}{G_{M-1}} \Big(\prod_{l=1, l\neq i}^{N} \rho_l^{n_l} \Big) \rho_i^{n_i-1} \quad (n_i > 0)$$

Thus, when a customer is transferred from i to j at time t, the situation it sees during its transfer (when it has left i but has not yet reached j) for the rest of the network is not $X(t-)$ but $X(t-) - e_i$. Therefore, the state of the network observed by this customer (and therefore not including it) is the same as the state of the network with $M - 1$ customers observed by an external observer at arbitary time.

$\square$

The product form solution for the stationary distribution also holds for other non-FIFO service disciplines. These issues were studied in a classic paper by Baskett, Chandy, Muntz and

Palacios called BCMP networks. They showed that a product form will still hold if we replace the FIFO assumption by LIFO (Last In First Out), Processor Sharing (PS), and infinite server models. The last two disciplines are of particular interest because they play an important role in the insensitivity of the stationary distribution to the assumption that the service times are exponential in the Jackson model. A natural question that can be posed is whether any of the assumptions, i.e., Poisson arrivals and exponential services, can be relaxed and under what conditions. Before doing so we study the notion of First-Order equivalents for Markovian networks as we saw in Chapter 2 where we defined the notion of first order equivalents through the notion of conditional arrival and departure rates. In the older literature this is referred to as Norton's theorem or equivalent in the case of closed queueing networks.

4.2.1   FIRST-ORDER EQUIVALENTS

As in Chapter 2 we can show that for Markovian qeueing networks a type of detailed balance equation can be written for each node in terms of the notion of a conditional intensity.

Recall for any stable queue with a stationary, ergodic arrival process $A_i(t)$ that possesses a $\mathcal{F}_t$ stochastic intensity $\lambda_i(t)$ where $\mathcal{F}_t$ is a suitable reference filtration, and corresponding departure process $D_i(t)$ whose $\mathcal{F}_t$-intensity is $\mu(t)$, the first-order equivalent is given by:

$$\lambda_n \pi(n) = \mu_n \pi(n+1), \quad n = 0, 1, 2, \ldots, \tag{4.21}$$

where $\pi(n) = \mathbb{P}(X(0) = n)$, $\lambda_n = \mathbb{E}[\lambda(0)|X(0) = n]$, and $\mu_i = \mathbb{E}[\mu(0)\mathbb{1}_{[X(0)>0]}|X(0) = n]$ respectively.

Let us see what this means in the context of a Jackson network.

Consider a network with N nodes with external Poisson arrival and service processes and routing probabilities P_{ij}. Let $X(t)$ denote the state of the Markov process where $X(t) = (X_1(t), X_2(t), \ldots, X_N(t))$ where $X_i(t)$ denotes the number of packets at queue or node $i = 1, 2, \ldots, N$.

Let $N_i(t)$ denote the external arrival at node i that is Poisson that possesses a a $\mathcal{F}_t^X$ stochastic intensity ν_i. Similarly, let $D_i(t)$ denote the counting process from node i and assume the process possesses a $\mathcal{F}_t^X$ intensity given by $\mu_i \mathbb{1}_{[X_i(t-)>0]}$, i.e., the services are i.i.d. exponential with mean $\frac{1}{\mu_i}$.

Then the total arrival process to node i when the state is $X(t)$ denoted by $A_i(t)$ is given by:

$$A_i(t) = N_i(t) + \sum_{j=1}^{N} \int_0^t P_{ji} d D_j(s), \quad i = 1, 2, \ldots, N \tag{4.22}$$

Note the intensity of the arrival process is thus $\nu_i + \sum_{j=1}^{N} \mu_j P_{ji} \mathbb{1}_{[X_j(0)>0]}$.

The departure process actually consists of those departures that exit from the network and those that go to other nodes and given by:

$$D_i(t) = D_{i0}(t) + \sum_{j=1}^{N} D_{ij}(t)$$ (4.23)

Hence, the conditional intensity

$$
\begin{aligned}
\lambda_i(n) &= \mathbb{E}[v_i + \sum_{j=1}^{N} \mu_j P_{ji} \mathbb{1}_{[X_j(0)>0]} | X_i(0) = n] \\
&= v_i + \sum_{j=1}^{N} \mu_j P_{ji} \mathbb{P}(X_i(0) > 0 | X_i(0) = n) \\
&= v_i + \sum_{j=1}^{N} P_{ji} \mu_j \rho_j \\
&= v_i + \sum_{j=1}^{N} P_{ji} \lambda_j \\
&= \lambda_i
\end{aligned}
$$

where λ_i is just the solution of the traffic equation given by (4.8).

By definition the conditional departure rate from node i is just μ_i and thus we see that the first order equivalent for node i in a Jackson network is just:

$$\lambda_i \pi_i(n) = \mu_i \pi_i(n+1), \quad n \geq 0$$ (4.24)

where:

$$\pi_i(n) = \sum_{\mathbf{x}:x_i=n} \pi(\mathbf{x})$$

and $\pi(\mathbf{x})$ is given by (4.10).

Remark 4.6 The first-order equivalent for a node can be directly extended to the case of Markovian networks with state dependent intensities. The only problem then is that one cannot characterize the conditional arrival and departure intensities as the stationary or steady-state distribution does not have an explicit form.

In the next section we address the question of when the steady-state distribution can be characterized in terms of mean arrival rates and service rates when we relax the assumption that the services are i.i.d. exponential. This is related to the notion of when a queueing network is *insensitive* to the service time distribution.

4.3   INSENSITIVITY IN STOCHASTIC NETWORKS

Insensitivity in stochastic networks refers to whether the stationary distribution can be characterized knowing a finite number of moments (typically the mean) of the service time distributions because usually the arrival processes are assumed to be Poisson for which the knowledge of the intensity or rate is necessary and sufficient to characterize the entire distribution. In this section we will study when this property holds for stochastic networks.

Let $\{X_t, \ t \in \mathbb{R}\}$ be a stationary, ergodic Markov chain on E with generator $q(x, y)$. Suppose that $\sup_x \sum_{y \neq x} q(x, y) < \infty$ and let $\pi(x)$ denote the stationary distribution. Then the global balance equation (GBE) can be written as:

$$\sum_y [\pi(y)q(y, x) - \pi(x)q(x, y)] = 0 \ \forall \, x \in E \tag{4.25}$$

Let us introduce the notation:

$$[A, B] \triangleq \sum_{x \in A} \sum_{y \in B} \pi(x)q(x, y) \tag{4.26}$$

to denote the probability flux from A to B that are subsets of E. Using this notation the GBE can be written as:

$$[E, x] = [x, E] \tag{4.27}$$

where $x \in E$ is the set of the singleton $x \in E$.

For some specific Markov Chains it happens that balance holds over a set $A \subset E$ in that:

$$[A, x] = [x, A] \ \ x \in A \tag{4.28}$$

Such a property is called *partial or local balance* over A. This amounts to the property that the equilibrium distribution is unchanged (modulo a normalizing factor) if transitions out of A are forbidden (as in the case of loss networks we have seen). By virtue of the GBE this can also be stated as:

$$[\bar{A}, x] = [x, \bar{A}], \ \ x \in A \tag{4.29}$$

If A is a singleton, then the property is called *detailed balance* as we saw with equilibrium properties of birth-death processes. Detailed balance over all pairs of states in E is equivalent to reversibility of the Markov process in equilibrium (as in the $M/M/1$ case). It turns out the partial balance is key to insensitivity whereby the equilibrium distribution $\pi(x)$ holds if we replace the exponential service times by i.i.d. generally distributed service times with the same mean. An interested reader should consult the works of Schassberger and the book of Kelly listed at the end of this chapter. So instead, we will see some special, and yet quite useful, network models that give rise to insensitive stationary distributions. A particularly nice set of models are so-called Whittle networks of which the so-called processor sharing (PS) server model and the $M/G/\infty$ model that underlies the loss network models are archetypes.

Consider a stochastic network of the Jackson type except that the service rate at each queue depends on the state x and given by $\mu_i \phi_i(x)$ for some non-negative function $\phi_i(x)$ with the convention $\phi_i(0, x_{-i}) = 0$ where x_{-i} denotes all components of x not including i. We call $\phi_i(x)$ the bandwidth allocation of node i in state x. Furthermore, it is assumed that $\sum_{i=1}^{N} \phi_i(\mathbf{x}) \leq 1$.

Definition 4.7   A stochastic network is said to be a Whittle type of network if the service capacities or allocations satisfy the following balance equation:

$$\phi_i(x)\phi_j(x - e_i) = \phi_j(x)\phi_i(x - e_j), \; \forall i, j \in \{1, 2, \cdots, N\}, \forall x : X_i > 0, x_j > 0 \qquad (4.30)$$

If the service requirements of packets at each queue are exponentially distributed with rate μ_i, and $\sum_{i=1}^{N} \phi_i(\mathbf{x}) = 1$, writing down the generator of the resulting Markov process shows that it will satisfy the Kolmogorov conditions given in the Appendix and is tantamount to the reversibility of the Markov process. In particular, the process satisfies the partial balance property and can also be shown to be insensitive. We can then state the following result.

Proposition 4.8   *Consider a network with N nodes and external arrivals arrive as a Poisson process with rate ν_i and the service times at queue i are i.i.d. with mean $\frac{1}{\mu_i}$. Let P denote the routing matrix and let $\lambda = (\lambda_1, \lambda_2, \cdots, \lambda_N)$ denote the solution of the traffic equation (4.7). Suppose the service capacities at each node satisfy the Whittle condition given in (4.30). Under the condition that $\lambda_i < \mu_i$, $i \in \{1, 2, \cdots, N\}$, there exists a unique stationary distribution given by:*

$$\pi(x) = \pi(0)\Phi(x) \prod_{i=1}^{N} \rho_i^{x_i} \qquad (4.31)$$

where $\Phi(.)$ is called the balance function and satisfies:

$$\phi_i(x) = \frac{\Phi(x - e_i)}{\Phi(x)}, \quad x_i > 0, i = 1, 2, \cdots, N \qquad (4.32)$$

Proof. The proof follows as in the proof of the Jackson network case. It is sufficient to show that the partial balance property is satisfied, which it indeed is because the detailed balance holds.   □

Remark 4.9   Markov processes satisfying the Whittle condition result in reversible processes but in order to show insensitivity we need to show that if the service times are approximated by hyperexponentials with the same mean, then the network is still a Whittle network. We cannot do this for a $M/M/1$ queue with FIFO and we have seen these queues are not insensitive to the service time distribution even though the queue is reversible.

Examples We now list some common models for which the insensitivity property holds and completely specify the stationary distribution.

1. Multi-class processor sharing (PS) queues

We can think of a multi-class processor sharing queue with class i arrival rates $v_i, i = 1, 2, \cdots, N$, and arrivals of type i requiring i.i.d. requiring a $\frac{1}{\mu_i}$ units of service with a server of unit capacity and trivial routing matrix where $p_{i0} = 1$. Let $x = (x_1, x_2, \cdots, x_N)$ denote the vector of the number of jobs of each type in the system. Under processor sharing each arriving job immediately goes into service and receives $\phi_i(x) = \frac{x_i}{x_1 + x_2 = \cdots + x_N}$.

It is easy to check the Whittle condition is satisfied and so to specify the stationary distribution we see $\lambda_i = v_i$ and is given by:

$$\pi(x) = \pi(0)\Phi(x) \prod_{i=1}^{N} \rho_i^{x_i}, \quad \rho_i = \frac{v_i}{\mu_i} \tag{4.33}$$

where the balance function is given by:

$$\Phi(x) = \frac{|x|!}{x_1! x_2! \cdots x_N!}, \quad x_1, x_2, \cdots, x_N > 0, |x| = \sum_{i=1}^{N} x_i \tag{4.34}$$

and the stability condition is $\sum \rho_i < 1$.

It is of interest to compute the mean sojourn time of a type i packet in the system denoted by W_i. This flows directly from Little's formula $\mathbb{E}[x_i] = v_i \mathbb{E}[W_i]$. Indeed:

$$
\begin{aligned}
\mathbb{E}[x_i] &= \sum_{x \in \mathbb{Z}_+^N, x_i > 0} x_i \pi(x) = \sum_{x : x_i > 0} \frac{|x|!}{x_1! x_2! x_i - 1! (x_i - 1)! x_{i+1}! \cdots x_N!} \\
&= \rho_i \sum_{x} (\sum_{j=1}^{N} x_j + 1) \pi(x)
\end{aligned}
$$

Suppose we would like to compute the average fraction of the sojourn time of a packet i in the system that the system is in a subset A of $\mathbb{Z}_+^N$. For example, we could take $A = \{x : \sum_{j=1}^{N} x_i \geq C\}$. For example, we can think of C as an occupancy level corresponding to a congestion event. Let us denote it by $F_i(A)$ defined as

$$F_i(A) \triangleq \frac{\mathbb{E}_i \left[\int_0^{W_i} \mathbb{1}_{\{X(t) \in A\}} dt \right]}{\mathbb{E}_i[W_i]}, \tag{4.35}$$

It follows from Little's formula and the Swiss Army formula of Palm calculus, cf. Chapter 2, respectively, that

$$\mathbb{E}_\pi[X_i] = \nu_i \mathbb{E}_i[W_i] \tag{4.36}$$

and

$$\mathbb{E}_\pi[\mathbb{1}_{\{X(0)\in\mathcal{A}\}} X_i(0)] = \nu_i \mathbb{E}_i\left[\int_0^{W_i} \mathbb{1}_{\{X(t)\in\mathcal{A}\}} dt\right]. \tag{4.37}$$

Therefore,

$$F_i(A) = \frac{\displaystyle\sum_{x\in\mathcal{A}} x_i \pi(x)}{\displaystyle\sum_x x_i \pi(x)}. \tag{4.38}$$

Remark 4.10   From the fact that the network satisfies the Whittle balance condition, when the service requirements are exponential, the Markov process is reversible and from the product-form of the distribution one can readily show that the departures of each class are Poisson with rate ν_i for class i. By then using the *method of stages* whereby general service requirements can be approximated by distributions with rational Laplace transforms, one can then show that the departure process in multi-class PS queues are mutually independent Poisson processes. We will use this property a bit later when we discuss Kelly networks.

The multi-class PS model is a very useful model to model the behavior of flows in a network where the average service requirement is thought of as the size of the flows as in the Internet. We will see a generalization of this model to the network case at the end of this section.

2. Jackson networks of infinite server queues

This another model that has been studied in the literature. The loss network model is a particular case of this model with fixed routes, i.e., $p_{ij} = 1$ if i and j are the neighboring nodes along a route and ν_i now refers to the arrival rate along route i. Here, instead, we consider the model as a variant of the Jackson network except that the server rate at node i is $x_i \mu_i$. Once again it can be trivially checked that the Whittle balance equation is satisfied trivially.

The stationary distribution is given by:

$$\pi(x) = \pi(0) \prod_{i=1}^N \frac{\rho_i^{x_i}}{x_i!} \tag{4.39}$$

where $\rho_i = \frac{\lambda_i}{\mu_i}$ and λ_i is obtained from the traffic equation, i.e. (4.7).

To see this one can check that $\Phi(x) = \frac{1}{\prod_{i=1}^N x_i!!}$.

We conclude our discussion of stochastic network models by considering another class of models called Kelly networks. A Kelly network is like a Jackson network with multiple classes of arrivals and the service discipline at each node being given by a multi-class allocation mechanism. In particular, we consider the case when the bandwidth allocation is according to the PS discipline based on the classes present at a node and the classes are defined by the routes. The case when the servers are infinite servers with a fixed total occupancy limit corresponds to the loss network model that we will not discuss.

Consider a network with N nodes and customer classes defined by the set of routes denoted by $\mathcal{R}$. A route is defined by a sequence of queues visited. Let the cardinality of $\mathcal{R}$ be R. A route $r \in \mathcal{R}$ is defined by the sequence of nodes $\{i_1, i_2, \cdots, i_r\}$ such that an arrival enters at i_1 and exits the network at node i_r. Let $A(r) = \{A_{i,r}\}$ denote the routing matrix with $A_{ir} = 1$ if node i is on route r and 0 otherwise. Packets arrive on route $r \in \mathcal{R}$ at rate v_r and we associate each route with a class. Packets of class k require $\frac{1}{\mu_{k,i}}$ units of service on average at node i and are i.i.d.

In a Kelly network, the state is the number of packets of each class in the network $x = (x_1, x_2, \cdots, x_R)$. Let us define $\rho_{k,i} = \frac{v_k}{\mu_{k,i}}$ as the average load brought by an arriving packet of class k, $k = 1, 2, \cdots, R$ at a node i along route k.

The average load on each node $i \in \{1, 2, \cdots, N\}$ is given by $\eta_i = \sum_{k=1}^{R} \rho_{k,i} A_{i,k}$. We can now state and we outline a proof for this type of Kelly network with PS at the nodes.

Proposition 4.11 Kelly Networks with PS

Consider a network of N processor sharing nodes with $\mathcal{R}$ routes of cardinality R. Let the arrivals on route $r \in \mathcal{R}$ be Poisson with rate v_r and the packets on a given route are i.i.d. and require an amount of service with mean $\frac{1}{\mu_{r,i}}$ at node i if $A_{ir} = 1$. Let $x = (x_1, x_2, \cdots, x_R)$ denote the vector of the number of types of packets in the network along each route and $\bar{x} = (\bar{x}_1, \bar{x}_2, \cdots, \bar{x}_N)$ denote the vector of the number of packets at each node in equilibrium.

If $\eta_i < 1$ at each node $i \in \{1, 2, \cdots, N\}$ then the network is stable and the stationary distributions of the number of packets along routes and at the nodes, denoted by $\pi(x)$ and $\pi(\bar{x})$ respectively, are given by:

$$\pi(x) \quad = \quad \pi(0) \prod_{i=1}^{N} \bar{x}_i! \prod_{j:A_{j,i}=1} \frac{\rho_{j,i}^{x_j}}{x_j!} \tag{4.40}$$

$$\pi(\bar{x}) \quad = \quad \prod_{i=1}^{N} (1 - \eta_i) \eta_i^{\bar{x}_i} \tag{4.41}$$

where $\rho_{j,i} = \frac{v_j}{\mu_{j,i}}$ denotes the average load of an arrival of class j at node i on its route, and $\eta_i = \sum_{j=1}^{R} \rho_{j,i} A_{ij}$ denotes the average load at the node.

Proof. The proof essentially follows from the fact that each node is a multi-class PS node whose stationary distribution at node j is given by:

$$\tilde{\pi}_j(x) = (\bar{x}_j)! \prod_{i:A_{ij}=1} \frac{\rho_{i,j}^{x_i}}{x_i!}, \quad \rho_{i,j} = \frac{\nu_i}{\mu_{i,j}}$$

Then from the independence of the nodes:

$$\pi(x) = \pi(0) \prod_{i=1}^{N} \tilde{\pi}_i(x)$$

and then $\pi(\bar{x}$ is obtained from the fact that $\bar{x}_i = \sum_j x_j A_{ij}$ and using the multinomial formula.
□

This result can also be adapted to infinite server nodes in the obvious way using the fact that the distribution of a multi-class infinite server queue is given by:

$$\pi_i(x) = \pi(0) \prod_{j:A_{i,j}=1} \frac{\rho_{j,i}^{n_j}}{n_j!} \tag{4.42}$$

With this we conclude our discussion of characterizing the stationary distributions of various stochastic models and we have shown some interesting properties. In the next section we study the problem of bandwidth allocation, choosing the appropriate $\phi_i(x)$ in the Whittle network model and we show that this is related to some deterministic optimization problems. Given this connection, one can implement balance functions with some desirable properties via on-line network optimization algorithms with the resulting stochastic network offering some robust statistical behavior such as insensitivity.

4.4   OPTIMIZATION AND BANDWIDTH ALLOCATION

The problem of bandwidth (or capacity) allocation is central to the sharing of server resources whether in multiprocessor systems, data centers, or networks. The bandwidth allocation strategy plays a crucial role not only in the stability region of a system but also the user level performance. The requirements of bandwidth sharing in stochastic systems has usually been addressed in the context of stability. The main criterion of any desirable bandwidth allocation strategy is that the stability region should be maximal or close to maximal. By maximal it means that if a load can be stabilized by some strategy then the stability region associated with the prescribed bandwidth sharing scheme should be as close as possible to the theoretical region under similar assumptions on the stochastic behavior of the load. However, one can ask the following question. Can we design a bandwidth allocation strategy that yields some desirable behavior of the underlying stochastic system? For example, we saw that if the bandwidth allocation scheme satisfies the Whittle balance property then the resulting network is insensitive. We can think of this as a form of stochastic robustness since performance predictions and calculations can be made without detailed statistical knowledge of the service requirements of jobs. In this section we discuss the issue of stability and robustness of bandwidth sharing strategies that are stable and result in insensitivity of the resulting stationary distribution for the network. We will show that indeed many of the interesting insensitive allocations are very close to the PS discipline that results in insensitivity and this in turn has very nice properties with respect to a deterministic optimization problem.

In the context of wireline networks with a fixed number of users or flows, bandwidth allocation has usually been studied as an optimization problem of suitable user utility functions over the constraint set specified by the stability region. Of course this depends on the way the server is shared.

Consider a stochastic network with J nodes and links with server rate (or capacity) $\{C_i\}_{i=1}^{J}$ accessed by randomly arriving flows of varying volumes or sizes. We consider a Kelly type model with $\mathcal{R}$ routes of cardinality R. An arriving flow is assigned a bandwidth that it keeps throughout its route. Only the amount of processing time changes depending on the other flows that share the nodes. We define a class by the route it takes and arrivals on route $r \in \mathcal{R}$ are Poisson rate λ_r, $r = 1, 2, \cdots, R$ and an arrival of type r brings work of amount σ_r that is i.i.d. with mean $\frac{1}{\mu_r}$. Let us define $\rho_r = \frac{\lambda_r}{\mu_r}$ as the average load brought by a type r arrival into the network. Let $A = \{A_{j,r}\}$ denote the incidence matrix with $A_{jr} = 1$ if route r uses node j and 0 otherwise. Let $\mathbf{x} = (x_1, x_2, \cdots, x_R)$ denote the vector of the number of flows of each class within the network and $\phi_i(\mathbf{x})$ denote the bandwidth allocated to class i, $i = 1, 2, \cdots, R$ in the network. If there are x_i flows of type i then each flow receives a service rate of $\frac{\phi_i(\mathbf{x})}{x_i} C_j$ at node j. We first define the notion of the capacity stability region for a given allocation that we assume is a closed, convex set in $\mathfrak{R}_+^R$ with a non-empty interior.

Definition 4.12   The capacity region of the network denoted by $\mathcal{C}$ is defined as:

$$\mathcal{C} = \{\mathbf{x} : \exists \{\phi(x)_i\}_{i=1}^{R} \; with \; 0 \leq \phi_i(x) \leq C : \text{the Markov process } \mathbf{X(t)} \text{ is ergodic}\} \quad (4.43)$$

where $\mathbf{X}(t) = (X_i(t), X_2(t), \ldots, X_R(t))$ where $X_i(t)$ denotes the number of flows of type i in the network at time t.

For the purposes of our exposition we will consider the capacity region as the convex region specified by the following link constraints:

$$\mathcal{C}_\phi = \{\mathbf{x} : \sum_r x_r A_{lr} \rho_r < C_l \; l =, 2, \cdots, J\} \quad (4.44)$$

here C_l denotes the capacity of resource or node l.

The choice of bandwidth allocation function $\phi_i(\mathbf{x})$ defines the statistical characteristics of the network. For example, if each node is FIFO independent of the class then from the results we have seen the statistical behavior not only requires knowledge of the average load but also higher order statistics of the flow volume. Moreover, there is no explicit characterization of the stationary distribution if it exists. We say that a bandwidth allocation is *maximally stable* if the given load corresponding to the flows can be stabilized or equivalently the resulting network is ergodic. The choice of the bandwidth allocation is usually made based on performance considerations and we first discuss the utility optimization framework that has become a very standard way of studying this issue.

Suppose the number of flows in the network is fixed, say $\mathbf{x} = (x_1, x_2, \cdots, x_R)$. Let $U_i(.)$, $i = 1, 2, \cdots, R$ denote the utility function associated with providing a unit of service to a flow of class i. For example, the utility function could be concave and increasing in the rate provided to flow i, or a function of the delay that results because of the bandwidth allocated to a flow of class i. Within a class we assume the bandwidth is shared equally. In particular, for the bandwidth allocation $\phi_i(\mathbf{x})$ the utility of a user is $U_i(\frac{\phi_i(\mathbf{x})}{x_i})$. Here we assume that the utility function is concave, increasing, and differentiable.

The utility optimization problem (referred to as network utility maximization (NUM)) corresponds to obtaining the allocation that solves the following problem:

$$\max_{\phi(.)} \sum_{j=1}^{R} x_j U_j \left(\frac{\phi_j(\mathbf{x})}{x_j} \right) \qquad (4.45)$$

subject to

$$\sum_{r=1}^{R} A_{l,r} \phi_r(\mathbf{x}) \leq C_l \quad l = 1, 2, \cdots, J \qquad (4.46)$$

Remark 4.13 In economics such an optimization problem involving the sums of user utilities is called a social optimization problem. The allocations $\{\phi_i(\mathbf{x})\}$ that maximize the sum are referred to as the social optimum allocations.

The choice of the utility function depends on the objectives. The boundary of R-dimensional space spanned by the utility functions over the constraint set corresponds to the Pareto optimal solutions. Pareto optimality means that it is not possible to improve all the utilities simultaneously, i.e., increasing the utility of a given flow will lead to the decrease in utility of at least one other flow. In general, the Pareto frontier as it is called is a $R - 1$ dimensional surface and thus there are infinitely many Pareto optimal points to choose from and there is no *best choice*. So other criteria are invoked to choose between the Pareto optimal points. One of the most common criteria is *fairness*. For example, the aim might be to equalize bandwidth allocation giving each class a certain fraction of the bandwidth. This can be wasteful of capacity when flows cannot use the bandwidth allocated to them. One of the most common utility functions chosen is the log utility function, i.e., $U_i(\mathbf{x}) = \ln(\mathbf{x})$. This utility function has some particularly interesting properties. The optimal allocation is called a Nash bargaining solution (NBS) or as if often referred to in the networking literature as *proportional fairness*. The NBS is the unique point on the Pareto surface that is scale invariant and satisfies the so-called Axioms of Fairness in terms of symmetry, etc., as termed by Nash. Other commonly used fairness criteria are max-min fairness that amounts to sharing bottleneck bandwidth equally amongst all flows using the bottleneck. Indeed, Mo and Walrand showed that it is possible to unify the different optimization criteria by considering (w, α)- fair allocations obtained by optimizing the parametric family of utility functions given by:

$$U_r(\mathbf{x}) = \begin{cases} \frac{w_r}{1-\alpha} x_r^{1-\alpha} & \text{if } \alpha \neq 1 \\ w_r \log(x_r) & \text{if } \alpha = 1 \end{cases} \tag{4.47}$$

The case $\alpha = 1$ with $w_r = 1$ corresponds to the NBS while in the limit as $\alpha \to \infty$ the allocations coincide with max-min fair solutions.

There is a vary rich theory associated with the bandwidth allocation for a fixed number of users $\mathbf{n}$. In particular, the dual problem related to the optimization problem given in (4.45) can be distributed at each node with the Lagrange multipliers corresponding to the link or server capacity constraints being obtainable from link level measurements only. We briefly discuss this issue.

Consider the modified utility functions defined by:

$$\tilde{U}_i(\mathbf{x}) = U_i(\mathbf{x}) - (\sum_{j=1}^{J} \mu_j A_{ji}) x_i, \quad i = 1, 2, \cdots, R \tag{4.48}$$

where the parameters $\mu_i, i = 1 = 1, 2, \cdots, J$ correspond to Lagrange multipliers obtained from the Karush-Kuhn-Tucker conditions associated with the link constraints:

$$\sum_{j=1}^{J} \mu_j (C_j - \sum_{r=1}^{R} \phi_r(\mathbf{x}) A_{jr} = 0 \quad i = 1, 2, \cdots, J \tag{4.49}$$

Then one can show that the Nash equilibrium for the modified utility functions defined as the equilibrium when no flow can improve its utility unilaterally corresponds to the social optimal for the original utilities. Formally, we define the Nash equilibrium as the point x^* satisfies:

$$x_i^* = \arg\max_{x_i} U_i(x_i, x_{-i}^*) \tag{4.50}$$

where x_{-i}^* denotes the optimal Nash equilibrium allocation for the other flows. The key point is one has to determine the Lagrange multipliers μ_i for each node and these are obtainable by simple distributed node level procedures requiring knowledge only of the flows using a given node and not all the flows in the network. However, we do not go into details here since our aim is to study the interplay between these bandwidth allocations and the stochastic behavior of the network.

To gain some insight let us consider the case of a single link with a log utility function for the link flows. Then it is easy to see that the solution to:

$$\max_{\phi(.)} \sum_{j=1}^{R} x_j \log \left(\frac{\phi_j(\mathbf{x})}{x_j} \right) \tag{4.51}$$

subject to

$$\sum_{r=1}^{R} \phi_r(\mathbf{x}) \leq C \tag{4.52}$$

is given by:

$$\phi_j(\mathbf{x}) = \frac{x_j}{\sum_{i=1}^{R} x_i} C \tag{4.53}$$

which corresponds to processor sharing. This demonstrates a social optimal solution for log utilities (i.e., proportionally fair) results a given in a *natural* robustness property in that the corresponding stochastic network is insensitive. A natural question is whether such an analysis extends to networks in a direct manner, i.e., do proportionally fair bandwidth allocations in network models also result in insensitivity? A more general question is if we can associate any balance function that satisfies Whittle's condition with a social optimal optimization problem?

Consider a network with server allocations $\phi_i(.)$, $i = 1, 2, \cdots, J$ satisfying (4.30), assume $\phi_i(.)$ is C_1 and assume that the allocations are scale invariant, i.e., $\phi_i(N\mathbf{x}) = \phi_i(\mathbf{x})$ for a scale factor N. Then under regularity conditions on $\phi_i(.)$ as functions of $\mathbf{x}$, the Whittle balance condition can be written as:

$$\frac{\frac{d}{dx_j}\phi_i(\mathbf{x})}{\phi_i(\mathbf{x})} = \frac{\frac{d}{dx_i}\phi_j(\mathbf{x})}{\phi_j(\mathbf{x})} \tag{4.54}$$

To see this define $f_i(\mathbf{x}) = \log(\phi_i(\mathbf{x}))$. Then from (4.30):

$$\frac{f_i(N\mathbf{x}) - f_i(N\mathbf{x} - e_j)}{\frac{1}{N}} = \frac{f_j(N\mathbf{x}) - f_j(N\mathbf{x} - e_i)}{\frac{1}{N}}$$

$$\frac{f_i(\mathbf{x}) - f_i(\mathbf{x} - e_j/N)}{\frac{1}{N}} = \frac{f_j(\mathbf{x}) - f_j(\mathbf{x} - e_i/N)}{\frac{1}{N}} \quad \text{by scale invariance}$$

$$\frac{d}{dx_j} f_i(\mathbf{x}) = \frac{d}{dx_i} f_j(\mathbf{x}) \quad \text{taking limits as } N \to \infty$$

$$\frac{\frac{d}{dx_j}\phi_i(\mathbf{x})}{\phi_i(\mathbf{x})} = \frac{\frac{d}{dx_i}\phi_j(\mathbf{x})}{\phi_j(\mathbf{x})}$$

It is easy to check that for a single link employing PS with R flows with $\phi_i(\mathbf{x}) = \frac{x_i}{\sum_{k=1}^{R} x_k}$ satisfies the above condition and moreover $\phi_i(N\mathbf{x}) = \phi_i(\mathbf{x})$ or $\phi_i(.)$ is scale invariant.

Using this characterization we can show the following result that we state as a proposition without proof since the analysis is quite delicate.

Proposition 4.14

Let $\phi_i(\mathbf{x})$ be a bandwidth allocation scheme that is asymptotically scale invariant, i.e., $\exists \tilde{\phi}_i(\mathbf{x})$ such that $\lim_{N \to \infty} \phi_N(\mathbf{x}) = \tilde{\phi}_i(\mathbf{x})$. Then $\tilde{\phi}_i(\mathbf{x})$ is a proportionally fair bandwidth allocation, i.e., $\tilde{\phi}(\mathbf{x})$ solves the social optimization problem (4.45) with a log utility function.

The above result shows that choosing a server bandwidth allocation judiciously such that the network belongs to the Whittle-Kelly class results in insensitivity of the stationary distribution of stochastic model with arriving and departing flows. This type of model can be used to model a network with fluid flows and we can exploit the insensitivity property for performance evaluation. We discuss this next.

4.5 MODEL FOR FLOW BASED ARCHITECTURES

We now show how insensitive models can be used to develop the Erlang equivalent for flow based systems. We begin by drawing upon the connection between processor sharing and the network utility optimization. We will then show how such models can be used in the context of flow based architectures to study congestion processes. Of importance to note is that there is a similarity in the stationary probability distribution to the multi-rate Erlang loss case seen in Chapter 3. We can exploit this structure to then obtain closed-form formulae for performace measures of interest.

For simplicity let us consider a single server of capacity C bits per second that is accessed by M types of flows. Flows (or sessions) of type $j \in \{1, 2, \ldots, M\}$ arrive as a Poisson process with rate v_k and the flow durations are i.i.d with mean σ_k. Consider the following bandwidth allocation problem: If capacity is available a flow of class k is allocated a desired rate r_k with $\max_k r_k < C$. Else during congestion, defined as the states in which the total desired capacity exceeds the available capacity, the capacity is shared according to processor sharing. This is a flow model where the rate r_k can be thought of as the rate of a coder, input link rate, etc., with the sources capable of altering their rate depending on the congestion as in TCP for example.

Let $\mathbf{x} = col(x_1, x_2, \ldots, x_M)$ denote the state of the system, where x_j is the number of flows of type j in equilibrium.

In this case, the rate received by a given source of type or class m is:

$$\phi_m(\mathbf{x}) \quad = \quad x_m r_m, \quad \sum_{j=1}^{M} x_j r_j \leq C \tag{4.55}$$

$$= \quad \frac{x_m}{\sum_{j=1}^{M} x_j} C, \quad \text{otherwise} \tag{4.56}$$

Defining the balance function $\Phi(\mathbf{x})$ as

$$\Phi(\mathbf{x}) \quad = \quad \prod_{m=1}^{M} \frac{1}{(x_m!) r_m^{x_m}} \quad \sum_{j=1}^{M} r_j x_j \leq C \tag{4.57}$$

$$= \quad \frac{|\mathbf{x}|!}{C^{|\mathbf{x}|} \prod_{m:x_m>0} x_m!} = \frac{1}{C} \sum_{m} \Phi(\mathbf{x} - e_m) \; if \; \sum_{j=1}^{M} x_j r_j > C \tag{4.58}$$

$$\tag{4.59}$$

Then we can check that:

$$\phi_m(\mathbf{x}) = \frac{\Phi(\mathbf{x} - e_m)}{\Phi(\mathbf{x})}, \quad m = 1, 2, \ldots, M$$

as required by the Whittle balance condition for insensitivity of the network.

Let $v_k \sigma_k = \alpha_k$ denote the average load brought by an arrival of a type k flow to the system. A necessary and sufficient condition for the existence of a stationary distribution, denoted by $\pi(\mathbf{x})$ is $\sum_{j=1}^{M} v_k \sigma_k < C$, and $\pi(\mathbf{x})$ is given by:

$$\pi(\mathbf{x}) = \pi(0)\Phi(\mathbf{x}) \prod_{m=1}^{M} \alpha_m^{x_m}. \tag{4.60}$$

Using the characterization of the balance function (4.57), the stationary distribution can be rewritten as

$$\pi(\mathbf{x}) = \begin{cases} \pi(\vec{0}) \displaystyle\prod_{m=1}^{M} \frac{\beta_m^{x_m}}{x_m!} & \text{if } \mathbf{x} \cdot \mathbf{r} \le C, \\[2em] \displaystyle\sum_{m=1}^{M} \rho_m \pi(\mathbf{x} - \mathbf{e}_m) & \text{Otherwise.} \end{cases} \tag{4.61}$$

where $\beta_m = \alpha_m / r_m$ is the normalized traffic intensity. The normalization constant, given by

$$G = \frac{1}{\pi(\vec{0})} = \sum_{\mathbf{x} \in \mathbb{Z}_+^M} \Phi(\mathbf{x}) \prod_{m=1}^{M} \alpha_m^{x_m}, \tag{4.62}$$

is finite if and only if $\rho < 1$, where ρ denotes the aggregate link load, i.e.,

$$\rho = \sum_{m=1}^{M} \rho_m = \sum_{m=1}^{M} \alpha_m / C. \tag{4.63}$$

As a sequel, the stability condition $\rho < 1$ will be assumed to be satisfied.

As mentioned above, the notion of congestion is the situation when the flows of type k receive less than r_k units of bandwidth. We will concern ourselves with evaluating two performance metrics: 1) The probability of congestion and 2) the time-average congestion rate that corresponds to the average fraction of its sojourn that a flow does not receive the desired rate r_k.

Let us first show some properties of the model.

Lemma 4.15 *For the flow model described above, the condition $\sum_{j=1}^{M} x_j r_j > C$ is equivalent to $\phi_m(\mathbf{x}) < x_m r_m$ for all classes that are present in the system, i.e., $x_m > 0, m = 1, 2, \ldots, M$*

Proof. That the condition $\phi_m(\mathbf{x}) < r_m x_m$ implies $\sum_{m=1}^{M} x_j r_j > C$ is immediate because of the PS assumption in congestion. To show that $\sum_{j=1}^{M} x_j r_j > C$ implies that $\phi_m(\mathbf{x}) < x_m r_m$ we need to show that $\Phi(x) > \frac{\Phi(x-e_m)}{r_m x_m}$ which can be readily shown by the definition of $\Phi(\mathbf{x})$. $\square$

Let $C_m = \{\mathbf{x} : \phi_m(\mathbf{x}) < r_m x_m\}$ be the set of states for which congestion occurs for flows of type $m \in \mathcal{M}$. The first congestion metric, the probability of congestion P_m, is defined in a straightforward manner,

$$P_m \triangleq \pi\left(X \in C_m\right). \tag{4.64}$$

There are two equivalent interpretations for P_m. It can either be seen as the long term average the flows of class-m are congested, or from the PASTA property, it is the steady-state probability that a flow of class-m enters a congested network.

The other congestion metric of interest, the time-average congestion rate, is a measure of the average fraction of time that an arrival does not receive its desired rate during its time in the system. Let τ_m be the sojourn time of class-m arrivals in the system. Define

$$F_m \triangleq \frac{\mathbb{E}_m\left[\int_0^{\tau_m} \mathbb{1}_{\{X(t) \in C_m\}} dt\right]}{\mathbb{E}_m[\tau_m]}, \tag{4.65}$$

where the expectation is taken with respect to the Palm measure for the point process of arrivals of class-m and $\mathbf{x}$ is the stationary state process. Then F_m denotes the ratio of the average time that a class-m flow spends in a congested state during its sojourn to the average sojourn time.

It follows from Little's formula and the Swiss Army formula of Palm calculus, cf. Chapter 2, respectively, that

$$\mathbb{E}_\pi[X_m] = \nu_m \mathbb{E}_m[\tau_m] \tag{4.66}$$

and

$$\mathbb{E}_\pi\left[\mathbb{1}_{\{X \in C_m\}} X_m\right] = \nu_m \mathbb{E}_m\left[\int_0^{\tau_m} \mathbb{1}_{\{X(t) \in C_m\}} dt\right]. \tag{4.67}$$

Therefore,

$$F_m = \frac{\displaystyle\sum_{\mathbf{x} \in C_m} x_m \pi(\mathbf{x})}{\displaystyle\sum_{\mathbf{x}} x_m \pi(\mathbf{x})}. \tag{4.68}$$

Although the congestion metrics can be evaluated directly, the calculation is extremely cumbersome for high capacity links or a large number of classes. By noting that the stationary distribution has a product-form we can exploit the techniques developed for loss systems in Chapter 3 to obtain accurate estimates. Let us see how this is done.

Calculation of Congestion Probabilities

It has been previously established that congestion will not occur for any class if the system state $\mathbf{x}$ satisfies the condition $\mathbf{x} \cdot \mathbf{r} \leq C$. Since, from Lemma 4.15, the states for which congestion occurs is the same for all flow classes, the probability of congestion will just be written as P. One can now write the probability of congestion as

$$P = \sum_{\mathbf{x}:\mathbf{x}\cdot\mathbf{r} > C} \pi(\mathbf{x}). \tag{4.69}$$

The following lemma shows that the expressions can actually be written as a function of far fewer states.

Lemma 4.16   *The probability of congestion can be written as*

$$P = \sum_{m=1}^{M} \frac{\rho_m B_m}{1 - \rho}, \tag{4.70}$$

with

$$B_m = \sum_{\mathbf{x}:C-r_m < \mathbf{x}\cdot\mathbf{r} \leq C} \pi(\mathbf{x}). \tag{4.71}$$

Proof. In view of (4.61),

$$P = \sum_{\mathbf{x}:\mathbf{x}\cdot\mathbf{r}>C} \pi(\mathbf{x}),$$

$$= \sum_{\mathbf{x}:\mathbf{x}\cdot\mathbf{r}>C} \sum_{m=1}^{M} \rho_m \pi(\mathbf{x} - \mathbf{e}_m),$$

$$= \sum_{m=1}^{M} \rho_m \left(\sum_{\mathbf{x}:\mathbf{x}\cdot\mathbf{r}>C} \pi(\mathbf{x}) + \sum_{\mathbf{x}:C-r_m < \mathbf{x}\cdot\mathbf{r} \leq C} \pi(\mathbf{x}) \right),$$

$$= \sum_{m=1}^{M} \rho_m (P + B_m).$$

Hence,

$$P = \sum_{m=1}^{M} \frac{\rho_m B_m}{1 - \rho}. \tag{4.72}$$

$\square$

Let π^B denote the stationary distribution for the multi-rate Erlang loss model given by (3.10) and (3.11). Noting that the stationary distributions π and π^B are proportional on those states $\mathbf{x}$ such that $\mathbf{x} \cdot \mathbf{r} \leq C$, it follows that

$$B_m = \frac{G^B}{G} P_m^B. \tag{4.73}$$

where P_m^B is the blocking probability for class m is an Erlang loss model.

Thus, the probability of congestion is:

$$P = \frac{G^B}{G} \sum_{m=1}^{M} \frac{\rho_m P_m^B}{1 - \rho}. \tag{4.74}$$

Although the above results completely specify the probability of congestion in terms of the blocking probability of a multirate Erlang loss system, the normalization constant G is different. However, when the system is large we can show that asymptotically the normalization constant for the balanced fair system and the Erlang loss system coincide. Moreover, from the explicit analytical expression for the approximation of the blocking probability we can obtain a closed form expression for the probability of congestion in large systems.

Let N denote the scaling parameter and we note that in the absence of congestion, class-i flows have independent, generally distributed durations with mean $\frac{1}{\mu_i} = \frac{\sigma_i}{r_i}$. In particular, the normalized traffic intensity $\beta_i = \alpha_i / r_i$ coincides with the corresponding parameter $\beta_i = \lambda_i / \mu_i$ of the loss system.

In view of results of Chapter 3, cf. Proposition 3.8 for the lightly loaded case (3.20), since $\rho < 1$, a tight approximation under large system scaling using the blocking probabilities P_m^B can now be established. It remains to calculate the normalization constant $G(N)$, which can be unwieldy. It turns out that one can show that $G^B(N) \approx G(N)$ for large N, where $G^B(N)$ and $G(N)$ denote the normalization constants of the loss and the flow-level models respectively.

Lemma 4.17

$$\frac{G^B(N)}{G(N)} \to 1 \quad when \ N \to \infty. \tag{4.75}$$

Moreover, the convergence is exponential in N.

With this insight, one arrives at the following conclusion.

Proposition 4.18   *Under the large system scaling,*

$$P(N) \sim \sum_{m=1}^{M} \frac{\rho_m P_m^B(N)}{1 - \rho}, \tag{4.76}$$

where

$$P_m^B(N) \sim e^{-NI} e^{\tau d \epsilon(N)} \frac{d}{\sqrt{2\pi N}\sigma} \frac{1 - e^{\tau r_m}}{1 - e^{\tau d}} \tag{4.77}$$

- *d is the greatest common divisor of $r_1, \ldots, r_M$,*

- *$\epsilon(N) = \frac{NC}{d} - \lfloor \frac{NC}{d} \rfloor$,*

- *τ is the unique solution to the equation $\displaystyle\sum_{m=1}^{M} r_m \beta_m e^{\tau r_m} = C$,*

- *$I = C\tau - \displaystyle\sum_{m=1}^{M} \beta_m \left(e^{\tau r_m} - 1\right)$,*

$$\bullet \; \sigma^2 = \sum_{m=1}^{M} r_m^2 \beta_m e^{\tau r_m}.$$

We omit the proof but it follows from the results for lightly loaded loss systems (3.20).

Time-Average Congestion Rates

Finally, the large system scaling will be applied to the time-average congestion rates (4.38).

The following lemma shows that the corresponding sums can be written as a function of far fewer states.

Lemma 4.19 *For all* $m, n = 1, \ldots, M$, *let*

$$Q_{m,n} = \sum_{\mathbf{x}: C - r_n < \mathbf{x} \cdot \mathbf{r} \leq C} x_m \pi (\mathbf{x})$$

and

$$Q_m = \sum_{\mathbf{x}: \mathbf{x} \cdot \mathbf{r} > C} x_m \pi (\mathbf{x}).$$

Then

$$Q_m = \frac{\rho_m P_m}{1 - \rho} + \sum_{n=1}^{M} \frac{\rho_n Q_{m,n}}{1 - \rho}. \tag{4.78}$$

Proof. By definition:

$$Q_m = \sum_{\mathbf{x}: \mathbf{x} \cdot \mathbf{r} > C} x_m \pi (\mathbf{x}),$$

$$= \sum_{\mathbf{x}: \mathbf{x} \cdot \mathbf{r} > C} x_m \sum_{n=1}^{M} \rho_n \pi (\mathbf{x} - \mathbf{e}_n),$$

$$= \sum_{n=1}^{M} \rho_n \sum_{\mathbf{x}: \mathbf{x} \cdot \mathbf{r} > C} x_m \pi (\mathbf{x} - \mathbf{e}_n),$$

$$= \sum_{n=1}^{M} \rho_n \sum_{\mathbf{x}: \mathbf{x} \cdot \mathbf{r} > C - r_n} (x_m + 1_{\{n=i\}}) \pi (\mathbf{x}),$$

$$= \rho_m P_m + \sum_{n=1}^{M} \rho_n (Q_m + Q_{m,n}),$$

from which the result follows. $\square$

Now let $P^B_{m,n}$ be the class-n blocking probability in a multirate loss system with capacity $C - r_m$. Then:

Proposition 4.20   *Under large system scaling,*

$$P^B_{m,n}(N) \sim e^{-N I_m} e^{\tau_m d \epsilon_m(N)} \frac{d}{\sqrt{2\pi N}\sigma_m} \frac{1 - e^{\tau_m r_n}}{1 - e^{\tau_m d}},$$

where

- d *is the greatest common divisor of* $r_1, \ldots, r_M$,

- $\epsilon_m(N) = \frac{NC - r_m}{d} - \left\lfloor \frac{NC - r_m}{d} \right\rfloor$,

- τ *is the unique solution to the equation* $\displaystyle\sum_{n=1}^{M} r_n \beta_n e^{\tau r_n} = C$,

- $\sigma^2 = \displaystyle\sum_{n=1}^{M} r_n^2 \beta_n e^{\tau r_n}$,

- $\tau_m = \tau - \frac{r_m}{N\sigma^2}$,

- $I_m = \left(C - \frac{r_m}{N}\right)\tau_m - \displaystyle\sum_{n=1}^{M} \beta_n \left(e^{\tau_m r_n} - 1\right)$,

- $\sigma_m^2 = \displaystyle\sum_{n=1}^{M} r_m^2 \beta_n e^{\tau_m r_n}$.

The following result, together with Theorem 4.18 and Proposition 4.20, provides the large system asymptotics of the time-average congestion rates.

Proposition 4.21   *Under large system scaling, for all* $m = 1, \ldots, M$:

$$F_m(N) \sim \frac{r_m}{NC(1 - \rho)} P_m(N) + \sum_{n=1}^{M} \frac{\rho_n}{1 - \rho} P^B_{m,n}(N). \tag{4.79}$$

We conclude with a presentation of a numerical comparison of the asymptotic formulae with the exact values for the congestion probability and time-average congestion probabilities of Theorems 4.18 and 4.21 for a single link case with $M = 3$ classes of traffic. The rate limits for flows are $r_1 = 1$, $r_2 = 1$, and $r_3 = 2$. As well the traffic intensities are $\alpha_1 = 1, \alpha_2 = 1$, and $\alpha_3 = 2$. We present results

Table 4.1: Congestion probabilities $\rho = 0.93$

N	Exact	Approximation
10	2.21e-1	2.50e-1
20	9.51e-2	1.07e-1
30	4.72e-2	5.38e-2
40	2.49e-2	2.81e-2

when link capacity was set to $C = 4.3$ corresponding to a heavy load of 0.93 but still in the stable region. The results are given in Table 4.1.

Similarly, we can compute the time-average congestion rates for each of the three classes corresponding to the average fraction of time during the sojourn of a flow of a given class when the system is in a congested state. We report these for $\rho = 0.6$. These are indicated in Table 4.2.

Table 4.2: Time-average congestion rates $\rho = 0.6$

N	Exact			Approximation		
	$F_1(N)$	$F_2(N)$	$F_3(N)$	$F_1(N)$	$F_2(N)$	$F_3(N)$
10	1.02e-03	1.29e-03	2.56e-03	1.02e-03	1.29e-03	2.56e-03
20	5.67e-06	7.10e-06	1.38e-05	5.67e-06	7.11e-06	1.38e-05
30	3.66e-08	4.57e-08	8.82e-08	3.66e-08	4.57e-08	8.82e-08
40	2.51e-10	3.13e-10	6.01e-10	2.51e-10	3.13e-10	6.02e-10
50	1.78e-12	2.21e-12	4.25e-12	1.78e-12	2.21e-12	4.25e-12

4.6   COMPLEX NETWORKS AND CHAOS

In the examples discussed above we have only been exposed to models where the stationary distribution has *product-form*. That is, the stationary distribution up to a function (the balance function) can be written as a product of Poisson-type or $M/M/1$ type marginals. However, most queueing network models do not behave in such a manner. For example, in many applications assuming Markovian routing probabilities p_{ij} is simply not justified. For example, to minimize delay and to balance loads in a network, upon completion of service a packet might be routed to a node with fewer packets than the others. This means that the routing probability is random depending on the number of packets in the downstream node. This is an example of Join the Shortest Queue (JSQ) policy. This makes the network now have state dependent routing and hence state dependent arrivals will destroy the product form stationary solution. However, in many cases when the number of nodes is large the interaction between the nodes results in a *mean field* phenomena as in statistical physics that results in asymptotic independence between queues that can substantially simplify the analysis and lead to an explicit computation of performance measures. We discuss this idea below through a very concrete randomized routing problem to show that not only is a large system amenable to explicit analysis but also exhibits very interesting probabilistic behavior.

Let us consider an example that was first studied by Vvedenskaya, Dobrushin and Karpelevich that is an archetype of the so-called *propagation of chaos* as it is termed. Loosely speaking a system of interacting entities exhibits *propagation of chaos* if they are initially statistically independent and any finite number of them evolve independently as the population size goes to infinity. Let us now formally describe the setting.

Consider a stochastic network of N identical servers of capacity C. Suppose jobs arrive at a rate $N\nu$ and each job is of size $\frac{1}{\mu}$ on average. Job sizes are assumed to be i.i.d. Upon arrival a job, L queues are picked at random and the job is routed to a server (out of the L picked servers) that has the smallest number of ongoing jobs. The server capacity is assumed to be shared equally amongst all jobs at the server. Let $(x_1^N, x_2^N, \cdots, x_N^N)$ denote the vector of the number of jobs at each server. Then it is easy to see that the necessary and sufficient condition for stability is $\rho = \frac{\nu}{C\mu} < 1$.

Now suppose instead of an arriving job being sent to the server with the smallest number of jobs, the arrivals were routed uniformly to a given server, i.e., the probability of being sent to server i, $i = 1, 2, \cdots, N$ is $\frac{1}{N}$. In that case, because of the Poisson thinning property we would have N independent processor sharing queues and the stationary distribution would be given by:

$$\pi^{PS}(x_1, x_2, \ldots, x_N) = (1 - \rho)^N \rho^{|\mathbf{x}|}, \quad |\mathbf{x}| = \sum_{i=1}^{N} x_i \qquad (4.80)$$

In other words, at any queue i, independently of the others, $\mathbb{P}(x_i \geq k) = \rho_i^{x_i}$ implying a geometrically decaying tail. Thus, if the queues start out independent they remain independent for all time. This is an easy case. Now let us consider the JSQ.

The routing probability for an arrival at time t denoted by $p_{0i}(t)$ is given by:

$$\begin{aligned} p_{0i}(t) &= 1 \quad x_i(t) < x_j(t), \ j \neq i, \ i, j \in \mathcal{L} \ |\mathcal{L}| = L \\ &= 0 \quad otherwise \end{aligned}$$

with ties being broken with equal probability.

Thus, we see the routing probability is state dependent and when N is finite the queues cannot be independent. Indeed, it is extremely difficult to compute the stationary joint distribution.

Let $r_k^N(t)$ be the proportion of servers with at least k jobs in progress at time t. Let $\pi_N(\mathbf{x})$ be the stationary distribution of the system where $\mathbf{x} = (x_1, x_2, \cdots, x_N)$ which exists if $\rho < 1$. Then the result of Vvedenskaya and Dobrushin is the following.

Proposition 4.22   *Consider a stochastic network of N identical servers with capacity C each. Consider jobs that arrive as a Poisson process with rate $N\nu$ and are exponentially distributed of mean size $\frac{1}{\mu}$. Upon* *arrival, L queues are picked uniformly at random and jobs are routed to the server (out of L) with the smallest number of on-going jobs. Let $\Pi_N(\mathbf{x})$ denote the stationary distribution. Suppose $\lim_{N \to \infty} r_k^N(0) \to u_0(k)$ in probability where $u_k(0)$ is a constant.*

Then for any finite number of L queues picked at random we have:

$$\lim_{N \to \infty} \mathbb{E}_{\Pi^N}(r_0(k)) = P_K = \rho^{\frac{L^k-1}{L-1}}, \quad k \in Z \tag{4.81}$$

or equivalently

Let $\Pi^{(L)}$ denote the restriction of Π^N to any L coordinates, with $\pi = \Pi^{(1)}$ being the one-dimensional marginal of Π. Then:

$$\Pi^{(L)} = \bigotimes_{i=1}^{L} \pi \tag{4.82}$$

In other words, there is asymptotic independence and moreover the tail distribution is super-exponential as opposed to simply being exponential as suggested by (4.80).

Although this result assumes a Markovian structure in the PS case it can be shown that queue distributions are insensitive to the job size distribution.

The importance of this result is that by choosing a JSQ policy but picking from a finite random set, the servers are highly *balanced* in that they will contain very few jobs on average. It can be seen that there is gain even when $L = 2$ is chosen, i.e., picking two servers at random and routing to the server with fewer jobs already gives the super-exponential decay. This is referred to as the *Power of Two* rule. It is instructive to see how the proof works and so below we provide a sketch for the case $L = 2$, assuming asymptotic independence is true and then deriving a result.

Sketch of Proof

Let us assume that from the chaos hypothesis that any two chosen queues are independent. First let us compute the average potential arrival rate to a given server, say 1. By the sampling assumption, the probability that the server is not chosen when picking two servers uniformly from N is $\frac{N-2}{N} = \frac{\binom{N-1}{2}}{\binom{N}{2}}$ and hence the probability of a potential arrival to the two queues is $(1 - \frac{N-2}{N}) = \frac{2}{N}$. In other words, the potential arrival rate to the servers is $N\nu\frac{2}{N} = 2\nu$. However, an arrival only enters a given server if it is the smaller of the two chosen.

Let π_k be the stationary probability that a server has k jobs and $P_k = \sum_{j=k}^{\infty} \pi_j$ denote the tail probability of having at least k jobs.

Then the probability that the arrival joins queue 1, by the independence hypothesis of the two queues, is $0.5 \times \mathbb{P}(second\ server = k) + 1 \times \mathbb{P}(second\ server > k) = 0.5\pi_k + P_{k+1}$ where π_k denotes the probability that the second server has k jobs too and P_{k+1} is the probability that the second server has more than k jobs by independence. Noting that $\pi_k = P_k - P_{k+1}$ we obtain the probability that an arrival joins server 1 is $0.5(P_k - P_{k+1}) + P_{K+1} =).5(P_k + P_{k+1})$. But the potential arrival rate is 2ν and hence the arrival rate to server 1 is $\nu_k = \nu(P_k + P_{k+1}$. From balance equations:

$$\nu_k \pi_k = \mu C \pi_{k+1}$$

and hence substituting for ν_k we obtain:

$$
\begin{aligned}
P_k &= \rho^{2^k-1} & (4.83)\\
\pi_k &= \rho^{2^k-1}(1-\rho^2) & (4.84)
\end{aligned}
$$

In a similar way if we take $L > 2$ then we obtain using similar arguments (except that there are more events that result in ties):

$$
\begin{aligned}
\lambda_k &= \lambda \frac{(P_k)^L - (P_{k+1})^L}{P_k - P_{k+1}}, & (4.85)\\
P_k &= \rho^{\frac{L^k-1}{L-1}} & (4.86)
\end{aligned}
$$

This asymptotic independence property holds for other network models of shared resources. For example, such an approach has been used to study the behavior of random access protocols in wireless networks where it is possible to provide rigorous justification of a heuristic due to Bianchi on the average throughputs of a contention-based protocol for medium access.

A key property that is required for chaos to occur is the property of exchangeability of the processes that roughly corresponds to the property that all permutations of the random processes have the same joint distributions.

4.7   STOCHASTIC FLUID NETWORKS

Stochastic fluid networks are another important class of network models especially in the context of flows where the rates are so high that the granularity of individual bits or packets does not matter. Except for the simple cases above explicit distributions for stochastic networks are not available and therefore we seek alternative methods to provide insights about their probabilistic behavior. In this section we first re-visit the Skorokhod problem that we briefly encountered in Chapter 2. The Skorokhod problem plays an important role in analyzing fluid stochastic network models. In particular, we will see some results on conditions for stability or existence of a stationary distribution and some results on monotonicity when we can bound the performance of stochastic networks by simpler and better characterized models.

The Skorokhod Oblique Reflection Problem (SP) states that given a càdlàg process $\{X(t) \in \mathfrak{R}_+^N; t \geq 0\}$ and an M-matrix[2] R, there exists a unique processes $\{W(t); t \geq 0\}$ (known as the reflected process) and $\{Z(t); t \geq 0\}$ (known as the regulator process) such that $\forall t \geq 0$:

1. $W(t) = X(t) + RZ(t) \geq 0$

2. $Z(0) = 0$ and $dZ(t) \geq 0$

3. $\int_0^t W_i(s)dZ_i(s) = 0, \quad i = 1, 2, \ldots, N$

Furthermore, there exists a unique, continuous pair of functions $\Phi, \Psi : D[0, \infty) \to D[0, \infty)$ such that $\Phi(X(\cdot)) = W(\cdot)$ and $\Psi(X(\cdot)) = Z(\cdot)$. A very useful fact is that if there exists another pair of processes $(\hat{W}, \hat{Z})$ that satisfies the first two properties, then $\hat{Z} \geq Z$. This is known as the minimality property of the regulator process.

Stochastic fluid networks are characterized by the 4-tuple $\{J, \mathbf{r}, P, W(0)\}$ where $\{J(t); t \geq 0\}$ is the cumulative input process, $\mathbf{r}$ is the service or release rate, P the routing matrix, and $W(0)$ is the initial workload. It is assumed that the queues are work conserving. We consider open stochastic fluid networks with N servers that have a single processor and infinite capacity buffers.

The cumulative inputs $\{J_i(t); t \geq 0\}, i \in 1 \ldots N$ are modeled as a non-decreasing, càdlàg processes with $J(0) = 0$. The routing matrix P is assumed to be sub-stochastic such that $(I - P^T)^{-1}$ exists and the spectral radius of P denoted by $sp(P) < 1$. The matrix $R = I - P^T$ is called the reflection matrix. By assumption, R belongs to the class of M matrices. The release rate $\mathbf{r} : \mathfrak{R}_+^N \to \mathfrak{R}_+^N$ is assumed to be a bounded, continuous, non-negative function.

Given the above primitives, the workload or reflected process for a stochastic fluid network is given by:

$$W(t) = W(0) + J(t) - \int_0^t (I - P^T)r(s)ds + (I - P^T)Z(t) \tag{4.87}$$

models the workload (or fluid buffer content) process of the network. Note here I represents the identity matrix.

We can now state the basic result on the stability properties of such a model.

Theorem 4.23 *Consider a stochastic fluid network with the properties above with constant service rate* $\mathbf{r}$ *and inputs that are process with stationary increments with* $\mathbb{E}[J(1)] = \lambda$ *and* $\text{var}(J(1)) = \sigma^2$. *Then under the condition:*

$$(I - P^T)^{-1}\lambda < \mathbf{r} \tag{4.88}$$

There exists a stationary distribution for the stochastic network.

---

[2]A M matrix is a matrix whose inverse is a positive matrix. Moreover, we also assume that the matrix R is also totally-S where an S matrix is one that maps non-negative vectors to non-negative vectors.

Note the above condition can also be written as $\lambda < (I - P^T)\mathbf{r}$.

Other than existence of a stationary distribution very little can be said about the moments. However, we can develop comparison theorems that allow us to bound the behavior of the network by a simpler network whose distributions are better characterized. Let us see some simple comparison results.

Consider a SFN with M nodes and an N-dim increasing Lévy input denoted by $J(t) = J(0, t)$ with $\lambda_i = \mathbb{E}[J_i(1)]$ where $J_i(t)$ is the i-th input process. Let $P = \{P_{ij}\}$ denote the routing matrix that is assumed to be sub-stochastic and with spectral radius $\sigma(P) < 1$. By assumption $R = I - P^T$ is a M-matrix. Let the release rate vector be denoted by $\mathbf{r}$. Let $W(t)$ be the N-dimensional vector of the workload or queue content process that satisfies the Skorokhod reflection problem given by:

$$W(t) = W(0) + J(t) - (I - P^T)rt + (I - P^T)Z(t) \tag{4.89}$$

where $Z(t)$ is the regulator process defined as the minimal process such that:

1. i) $Z_i(t) \geq 0$ and continuous and increasing in t.

2. ii) $\int_0^t W_i(s)dZ_i(s) = 0, \quad i = 1, 2, \ldots, N$

It is easy to see that if $J(t)$ and $H(t)$ are two input processes with $J(t) \leq H(t) \;\; \forall t > 0$ then if $W^1(t)$ and $W^2(t)$ denote the corresponding workload processes, if $W^1(0) = W^2(0)$, then $W^1(t) \leq W^2(t) \; \forall t \geq 0$. In a similar way if $r_1 \leq r_2$ then $W^1(t) \geq W^2(t)$. In other words, the queues are monotonic in the inputs and release or service rates. A natural question is if we change the routing matrices what sort of monotonicity can we expect? It turns out that result depends on the assumptions.

First note that $R = (I - P^T)$ determines how the boundary reflections affect the individual coordinates or workloads. R is referred to as the reflection matrix. In particular, it is the columns of R that determine how fast a given process $W_i(t)$ is *pushed* away from the boundary when it hits 0.

We now state and prove the main comparison result that is useful in many applications.

Proposition 4.24   *Consider the stochastic fluid network model given by (4.89) with input J, release rate $\mathbf{r}$, and reflection matrix $R = (1 - P^T)$ denoted by $\{J, \mathbf{r}, P, W(0)\}$. We assume that $R = I - P^T$ is a M-matrix and totally S. Then the following result holds pathwise:*

Let $W_1(t)$ and $W_2(t)$ be the workloads corresponding to the model $\{J, \mathbf{r}, P_i), W_i(0)\}, i = 1, 2$ for the same input process $J(.)$, the same release rates $\mathbf{r}$ but with routing matrices P_1 and P_2 respectively. If $P_1 < P_2$,[3] and $W_1(0) \leq W_2(0)$, then $W_1(t) \leq W_2(t)$.

In particular, if the networks are stable, $(I - P_i^T)^{-1}\lambda < \mathbf{r}, i = 1.2,$, then for all $j = 1, 2, \cdots, N$

$$\mathbb{P}(W_{1,j} > x) \leq \mathbb{P}(W_{2,j} > x) \; x \geq 0 \tag{4.90}$$

where $W_{i,j}$ is the $j - th$ component of W_i.

---

[3] $P_1 < P_2$ means $P_2 - P_1$ is a P-matrix (with all nonnegative entries).

Proof. The proof essentially follows from the following monotonicity property: Consider two networks $\{J, \mathbf{r}_1, P, W(0)\}$ and $\{J, \mathbf{r}_2, P, W(0)\}$. If $r_1 < r_2$ and then $W_1(t) \geq W_2(t); \; t \geq 0$ where

$$W_i(t) = J(t) - (I - P^T)\mathbf{r}_1 t + (I - P^T)Z_i(t)$$

The monotonicity just follows by noting that $X_1(t) = J(t) - \mathbf{r}_1 t \geq X_2(t) = J(t) - \mathbf{r}_2 t$ and the definition of the reflection map.

Now by definition $(J, \mathbf{r}, P_1, W_1(0))$ and $(J, \mathbf{r}, P_2, W_2(0))$:

$$
\begin{aligned}
W_1(t) &= W_1(0) + J(t) - (I - P_1^T)\mathbf{r}t + (I - P_1^T)Z_1(t) \\
W_2(t) &= W_2(0) + J(t) - (I - P_2^T)\mathbf{r}t + (I - P_2^T)Z_2(t)
\end{aligned}
\tag{4.91}
$$

Now if $P_1 \leq P_2$ it readily follows that $(I - P_1^T)\mathbf{r}t \geq (I - P_2^T)\mathbf{r}t$ and hence $X_1(t) = J(t) - (I - P_1^T)\mathbf{r}t \leq X_2(t) = J(t) - (I - P - 2^T)\mathbf{r}t; \; t \geq 0$.

Now we can re-write (4.91) as:

$$W_1(t) = W_1(0) + J(t) - (I - P_1^T)\mathbf{r}t + (I - P_2^T)Z_3(t)$$

where $Z_3(t) = (I - P_2^T)^{-1}(I - P_1^T)Z_1(t)$.

Invoking the monotonicity result above, for $W_1(0) = W_2(0)$, we directly obtain:

$$W_1(t) \leq W_2(t) \quad t \geq 0$$

The stochastic dominance follows from the pathwise dominance, and under stationary assumptions the second result follows. $\qquad\square$

Remark 4.25 The above ideas can also be used to establish the following pathwise for the regulator processes $Z_i(t), i = 1, 2$ using the fact that they are minimal.

Under the conditions of the above theorem for every $0 < s < t$:

$$Z_1(t) - Z_1(s) \geq Z_2(t) - Z_2(s) \tag{4.92}$$

Using the above comparison result we can bound the mean workload in stochastic fluid networks by some simple network dominations.

Consider the following three stochastic networks :

a) $W(t) = J(t) - (I - P^T)rt + (I - P^T)Z(t)$

b) $W^u(t) = J(t) - (I - P^T)rt + Z^u(t)$

c) $W^l(t) = J(t) - rt + Z^l(t)$

Theorem 4.26 *Consider a stochastic fluid network whose input $J(t)$ is a stationary increasing process with $\mathbb{E}[J(1)] = \lambda$.*

Let $(I - P^T)^{-1}\lambda < r$. Then there exist stationary versions of the three networks and for every t we have the following pathwise result (interpreted component wise).

$$W^l(t) \leq W(t) \leq W^u(t), \quad t \geq 0 \text{ and } W(0) = W^u(0) = W^l(0) \tag{4.93}$$

Using this we can compute bounds when the input is a multi-dimensional Lévy process with independent components. This is because the process $W^l(t)$ is just a network of N independent queues with Lévy inputs (fluid $M/G/1$ queues) while $W^u(.)$ also can be viewed as a network of N independent fluid queues. The mean waiting time is then just given by the Pollaczek-Khinchine formula that we developed in Chapter 1. We state the result below.

Proposition 4.27 *Let $J(t) = \int_0^t Y_s dN_s$ where $\{N_t\}$ is is Poisson process with rate v and $Y_t = \Delta J_t$ and the jumps are independent of N. Define $v\mathbb{E}[Y_0] = \lambda$ and $var[Y_0^2] = \sigma^2$.*

Suppose that $\lambda < (I - P')r$, then

$$\frac{\sigma^2}{2(r - \lambda)} \leq E[W] \leq \frac{\sigma^2}{2((I - P^T)r - \lambda)} \tag{4.94}$$

interpreted component wise.

There are many more interesting results such as comparison theorems when the release rates and routing probabilities are state dependent, and the study of such networks in heavy traffic that continues to be a very active area of research. Characterizing end-to-end properties such as delays in stochastic networks continues to be a very challenging problem.

CONCLUDING REMARKS

In this chapter we have seen some models of stochastic networks where we can characterize the stationary behavior explicitly. In addition we have seen some simple properties that even though there is a type of independence among node occupancies, the internal flows are not Poisson even when external inputs are Poisson and the service requirements are exponential. Stochastic network models that we have seen in this chapter are useful in many applications. However, have seen that explicit characterizations of their stationary behavior is only possible under restricted assumptions. The most common are Markovian models popularized in the works of Jackson, BCMP et al., Kelly, and of course the seminal work of Whittle on insensitivity. As we have seen, especially when networks are insensitive to holding time or service time distributions,we can construct very useful models for flow architectures that mimic the Erlang theory for loss or circuit switched systems. Moreover, they are amenable for performance calculations. We have also seen the nice connection between insensitivity and the network optimization problems. This suggests that in congestion control algorithms that can be related to an optimization problem for log utilities the resulting bandwidth allocations make the network insensitive in stationarity suggesting that flow type models discussed here could provide good means to dimension router capacity when used with TCP type mechanisms.

The analysis of general stochastic models remains a very open question. The question of estimating end-to-end delays in networks remains a challenging one and only approximations are available once we leave the realm of Markovian networks. However, as shown in the fluid network case, some insights could be derived when comparison theorems are available.

NOTES AND PROBING FURTHER

BOOK REFERENCES

There are many excellent books that deal with stochastic networks at varying levels of generality. Below are a few excellent references that any interested reader should consult.

F. P. Kelly: *Reversibility and Stochastic Networks*, J. Wiley and Sons., Chichester, 1979.

> This is an excellent classic reference for understanding general Markovian models of networks, reversibility, and insensitivity.

P. Whittle, *Systems in stochastic equilibrium*, Wiley Series in Probability and Mathematical Statistics: Applied Probability and Statistics. John Wiley & Sons, Ltd., Chichester, 1986.

> This is a very interesting and informative monograph on stochastic systems, primary networks, in equilibrium. The issues of balance, insensitivity, and other properties are extremely well motivated and presented. The wide applicability of network models in various applications such as telecommunications, epidemics, population models, etc., are presented along with the background theory of Markov processes.

R. Serfozo, *Introduction to Stochastic Networks*, Springer, Berlin Heidelberg New York, 2005.

> This is a very comprehensive book and modern on stochastic networks. In has a very good treatment of flows in networks, the insensitivity property, and the stability analysis of both fluid and networks with general stationary inputs.

P. Robert, *Stochastic Networks and Queues*, Springer, Applications in Mathematics 52, Springer-Verlag, Berlin, 2003.

> This book is also excellent for learning the mathematical modeling and analysis of networks. The book presents a very comprehensive treatment of network processes and is particularly useful for results on point processes and their functionals. A very nice treatment of the Skorokhod

reflection problem is also given. It is a highly recommended book for the mathematically inclined.

H.C. Chen and D.D. Yao, *Fundamentals of Queueing Networks: Performance, Asymptotics, and Optimization*, Applications of Mathematics 46, Springer-Verlag, NY, 2001.

This is a very comprehensive view of queueing networks from the fluid viewpoint. The primary focus of this book is on stability theory, heavy traffic limit theorems, and on scheduling. Of particular note is the treatment of the reflection map associated with the multi-dimensional Skorokhod reflection problem.

W. Whitt, *Stochastic-process limits*, Springer-Verlag, NY, 2002.

This book is a very useful book for understanding fluid models of stochastic networks particularly understanding the multidimensional reflection maps that occur in the context of the Skorokhod problem. It is also a good reference for the analysis techniques such as the appropriate spaces and topologies to study fluid and diffusion limits.

S. Asmussen, *Applied probability and queues*, Applications of Mathematics 51, Springer-Verlag, N.Y., 2003

This book presents a necessary background in applied probability and stochastic processes for studying general queueing models. It is an excellent reference for the mathematically inclined and curious.

F. Baccelli and P. Brémaud, *Elements of Queueing Theory: Palm–Martingale Calculus and Stochastic Recurrences*, 2nd Ed., Applications of Mathematics 26, Springer-Verlag, NY, 2005.

In addition to Palm theory this is an excellent reference on stochastic networks with general stationary inputs and for the wealth of results on stochastic majorization and comparison of queues. A detailed presentation of Markovian networks and the properties of flows within can also be found in the book by Brémaud below.

P. Brémaud, *Point processes and queues: Martingale dynamics*, Springer-Verlag, NY, 1981.

M. Bramson, *Stability of Queueing Networks*, Lecture Notes in Mathematics Vol 1950, Springer, Berlin, 2008.

This is an excellent reference for learning techniques to study the stability issues in queueing networks. Of particular note is that the book thoroughly discusses the fluid limit approach and the instances when the *usual* $\rho < 1$ condition fails in some network models.

JOURNAL ARTICLES

The area of stochastic networks is a very active one and there are a number of excellent papers that appear every year on various aspects of network models. The papers listed below correspond to some of the classical papers as well as some recent ones where the connections between insensitivity, network optimization, pricing, etc., are given.

F. Baskett, M. K. Chandy, R. R. Muntz, and F. G. Palacios, Open, closed, and mixed networks of queues with different classes of customers, *J. Assoc. Comput. Mach.* 22 (1975), 248–260.

This is a seminal paper on Markovian networks. The results are usually called BCMP networks and provides an exhaustive list of results on conditions when Markovian networks exhibit product form stationary distributions.

B. Melamed, Characterizations of Poisson traffic streams in Jackson queueing networks, *Adv. in Appl. Probab.* 11, no. 2 (1979): 422–438.

B. Melamed, On the reversibility of queueing networks, *Stochastic Process. Appl.* 13, no. 2 (1982): 227–234.

The papers listed above due to Melamed study traffic flows in Markovian networks. The key results relate when flows are Poisson and the so-called "job observer" property. In particular, the non-Poissonian nature of internal flows is discussed in detail.

M. T. Hsiao and A. A. Lazar, An extension to Norton's equivalent, *Queue. Syst. (QUESTA)* 5 (1989): 401–412.

This paper discusses the notion of first-order equivalents for multi-class Markovian queueing networks in equilibrium. They show that the first-order equivalent is just a generalization of the notion of Norton's equivalent for closed Markovian networks.

R. Schassberger, The insensitivity of stationary probabilities in networks of queues. *Adv. in Appl. Probab.* 10, no. 4 (1978): 906–912. Correction: *Adv. in Appl. Probab.* 10, no. 4 (1978): 906–912.

This paper is a precursor to the papers of Whittle on the insensitivity property of certain stochastic network to job holding times. The approach is via the GSMPs (Generalized Semi-Markov Processes) that result when looking at the embedded chains.

P. Whittle, Partial balance and insensitivity, *J. Appl. Probab.* 22, no. 1 (1985): 168–176.

P. Whittle, Partial balance, insensitivity and weak coupling, *Adv. Appl. Probab.* 18, no. 3 (1986): 706–723.

The two papers of Whittle deal with partial balance and insensitivity and also consider the GSMP approach.

T. Bonald and A. Proutière, Insensitive bandwidth sharing in data networks, *Queue. Syst. Theory Appl.* 44, no. 1 (2003): 69–100.

T. Bonald, A. Proutière, J. Roberts, and J. Virtamo, Computational aspects of balanced fairness, in *Proceedings of 18th International Teletraffic Congress*, Elsevier Science, 2003, 801–810.

These two papers discuss the insensitivity property of networks and introduce the notion of balanced fairness that results in the networks satisfying the Whittle conditions. Balanced fairness coincides with proportional fairness in many topologies of interest and has a maximal property in that the bandwidth allocations are Pareto optimal.

L. Massoulié, Structural properties of proportional fairness: Stability and insensitivity, *Ann. Appl. Probab.* 17, no. 3 (2007): 809–839.

N. Walton, Insensitive, maximum stable allocations converge to proportional fairness, *Queue. Syst.* 68 (2011): 51–60.

The papers of Massoulie and Walton are the first papers to systematically study the relationship between network utility optimization and insensitive network bandwidth allocation. In particular, they establish that the rate functions associated with the tail distributions correspond to the allocations that result from proportionally fair allocations obtained from optimizing log utility functions. Walton extends this to so-called max stable allocations and shows that in the fluid limit the allocations result in proportional fair allocations.

F.P. Kelly, Charging and rate control for elastic traffic, *Euro. Trans. on Telecommun.* 8 (1997): 33–37.

J.W. Roberts and L. Massoulié, Bandwidth sharing and admission control for elastic traffic. *Telecommunication Systems*, 15 (2000): 185–201.

H. Yaiche, R. R. Mazumdar, and C. P. Rosenberg, A game theoretic framework for bandwidth allocation and pricing in broadband networks, *IEEE/ACM Trans. Network.*, 8, no. 5 (2000): 667–678.

J. Mo and J. Walrand, Fair End-to-End Window-Based Congestion Control, *IEEE/ACM Trans. Network.* 8 (2000): 556–567.

These four papers present the framework for the optimization approach to congestion and bandwidth allocation in networks. The approach is a deterministic one with fixed number of users. The paper of Kelly introduces the notion of proportional fairness and the system, network, and user optimization problems. The paper of Massoulié and Roberts shows how to view TCP-like schemes as a utility optimization problem with log utilities. The third paper shows the connection between the different optimization frameworks and the corresponding game theoretic solutions as well as introduces the primal-dual implementation framework. The last paper defines a general class of utilities from which the various known measures of fairness such as maxmin fair, proportional fairness, etc., follow as special cases. This is sometimes referred to α fairness.

J-P. Haddad and R. R. Mazumdar, Congestion in large balanced fair systems, *Queue. Syst. (QUESTA)*, Special Issue on Network Asymptotics, August 2012, 36 pages. DOI 10.1007/s11134-012-9322-x.

This paper studies the large system calculations in large balanced fair networks. The case of a single link presented in the chapter for flow-based architectures is taken from this paper.

A.D. Skorokhod, Stochastic equations for diffusions in a bounded region, *Theory of Prob. and its Appl.*, 6 (1961): 264–274.

This is the classic paper of Skorokhod on reflected processes that forms the basis for analyzing and studying fluid stochastic networks.

O.Kella, O. and W. Whitt, Stability and structural properties of stochastic storage networks, *J. Appl. Probab.*, 33, no. 4 (1996): 1169–1180.

This paper is a very important reference for stability results and moments in stochastic fluid networks that are referred to as storage networks.

S. Ramasubramanian, A subsidy-surplus model and the Sskorokhod problem in an orthant, *Math. Oper. Res.*, 25, no. 3 (2000): 509–538.

This paper studies comparison results for the multidimensional Skorokhod problem in the positive orthant.

J.P. Haddad, R. R. Mazumdar, and F. J. Piera, Pathwise comparison theorems for stochastic fluid networks,*Queue. Syst. (QUESTA)*, 66 (2010): 155–168.

This paper presents comprehensive comparison results for stochastic fluid networks, some of which are discussed in the last section of this chapter to obtain bounds on mean workload in queues.

N.D. Vvedenskaya, R.L. Dobrushin, and F. I Karpelevich, A queueing system with a choice of the shorter of two queues—an asymptotic approach, *(Russian) Problemy Peredachi Informatsii* 32, no. 1 (1996): 20–34; translation in *Probl. Inform. Transmis.* 32, no. 1 (1996): 15–27.

C. Graham, Chaoticity on path space for a queueing network with selection of the shortest queue among several,*J. Appl. Probab.* 37, no. 1 (2000): 198–211.

The above two papers present the analysis of the Join the Shortest Queue (JSQ) problem that has been discussed in this chapter.

G. Bianchi, Performance analysis of the IEEE 802.11 distributed coordination function, *IEEE J. Selected Areas Commun.*, 18, no. 3: 535–547.

C. Bordenave, D. McDonald, and A. Proutière, A particle system in interaction with a rapidly varying environment: mean field limits and applications, *Netw. Heterog. Media* 5, no. 1 (2010), 31–62.

The first paper derives the throughput in a contention-based access protocol for wireless networks based on a mean field approximation. The second paper provides a rigorous approach for using mean field and propagation of chaos idea to analyze MAC protocols.

M. Mitzenmacher, A. W. Richa, and R. Sitaraman, The power of two random choices: a survey of techniques and results, *Handbook of randomized computing, Vol. I, II*, 255–312, Comb. Optim., 9, Kluwer Academic Publishers, Dordrecht, 2001.

This paper presents a survey of results on the Power of Two choices in a number of applications.

Graham, Carl Chaoticity for multiclass systems and exchangeability within classes, *J. Appl. Probab.* 45, no. 4 (2008): 1196–1203.

This final paper is on the role of exchangeability in the existence of chaos in interacting systems.

CHAPTER 5

Statistical Multiplexing

INTRODUCTION

Statistical multiplexing refers to the phenomenon whereby sources with statistically varying rates are mixed together into a common server or buffer. Typical sources related to real-time phenomena such as speech or video are *bursty*—there are periods when they generate bits or packets at a high rate (ON state) while there are other periods when they generate a few or no packets (OFF state). Typical source behavior is depicted in the Figure 5.1:

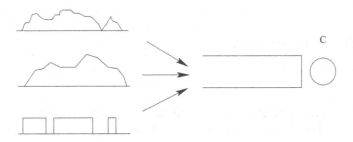

Figure 5.1: Statistical multiplexing of bursty sources.

Because of statistical independence between different users or sources it is a very unlikely scenario that all sources will be simultaneously in the ON state (especially when there are many), and thus to design a server to serve at a rate corresponding to the maximum sum rate of all the sources would be very wasteful. If we allow for a fraction of the offered traffic to be lost then we will see that it is possible that a link of given capacity can carry much more traffic than would be the case if we assumed that the traffic was synchronized.

A caricature of a typical scenario whereby one could exploit the statistical independence is shown in Figure 5.2:

In this Chapter we will study the issue of statistical multiplexing with an aim to quantify the gains. Moreover, we will see that there is a very important concept that emerges, namely the notion of *effective bandwidths*. This has been one of the major conceptual advances that emerged in the 1990s when the issue of providing Quality of Service (QoS) became important in ATM networks. The attractiveness of this idea is that it allows us to map a packet level phenomenon that can include queueing to a flow or connection level phenomenon i.e., allows us to ignore a queueing model and convert it into a loss model. This idea will be made precise in the sequel.

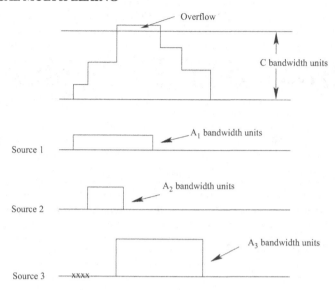

Figure 5.2: How statistical multiplexing works.

Quality of Service (QoS) is a buzzword that became popular in the late 1980s when the idea of Asynchronous Transfer Mode (ATM) networks emerged. The idea here was that the transport of packets or bits within a network was going to be assured by an architecture that resembled a circuit-multiplexed scenario (fixed paths) except that rather than reserving resources along a path fixed at a given level, the traffic streams were to be split into packets of fixed size called cells and the statistical variation in cell generation as in Figure 5.1 was to be exploited to "pack in" many more connections by allowing the variability in rates to be used. However, since one could not predict the OFF moments there would be periods when there would be temporary overload of the capacity and cells would be lost as in Figure 5.2. So, if one could tolerate the loss of some bits (not critical in many voice and video applications) then one could pack in many more sources otherwise in the situation shown in Figure 5.2 source 3 would be blocked. In the ATM context probability of packet (cell) loss was set to be of the order of 10^{-6}. This is indeed a very small probability and thus the question arising is how does one determine that indeed such a criterion is being met. Simulations are one way but that would be extremely cumbersome given the extremely small probabilities that need to be estimated. Thus, arose a need to develop methods to estimate quantities like cell or packet loss of very small magnitude. We will discuss these issues later in the chapter. We first begin by defining the various QoS metrics of interest.

5.1 PERFORMANCE METRICS FOR QUALITY OF SERVICE (QOS)

The key performance metrics in a network are the notions of average throughput, delay characteristics and loss. In the context of statistical multiplexing we will restrict ourselves to two of these measures namely delay and packet or bit loss.

In the context of delays, the key statistical measures are the average delay and the delay distribution. QoS requirements are usually specified by bounds on the average delay or by giving bounds on the probability of the delay exceeding a certain level. Thus, if D represents the delay, the performance measures are usually $\mathbb{E}_A[D] < D_{\max}$ or the $\Pr(D > D_{\max}) \leq \varepsilon$ where ε is a very small number. Here $E_A[.]$ denotes the mean with respect to the arrival distribution.

In the case of packet or bit loss we are usually interested in $\Pr(\text{ Packet (bit) is lost}) \leq \epsilon$. We will see that both the packet loss probability or the delay distribution are related to computing the tail probabilities associated with buffer occupancy or capacity exceedence.

Let us first define the various quantities. We begin with a simple scenario when there is no buffering. Figure 5.3 shows a typical sample path of the congestion and overflow process assuming sessions with different but fixed transmission rates arriving at a link.

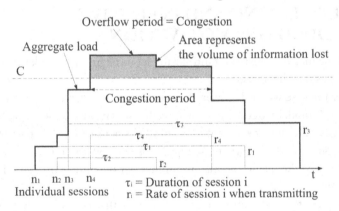

Figure 5.3: Congestion and overflow process.

Let X_t denote the instantaneous rate of a source transmitting on a link of capacity C bits per second. Assume $\mathbb{E}[X] < C$. Then the total number of bits lost in an interval of length T is: $Bits\ lost = \int_0^T (X_t - C)^+ dt$. If the source is stationary and ergodic then the average number of bits lost is

$$Loss = \lim_{T \to \infty} \frac{1}{T} \int_0^T (X_t - C)^+ dt = \mathbb{E}[(X_0 - C)^+] \tag{5.1}$$

$$= \int_C^\infty (1 - F(x)) dx$$

$$\approx const.\mathbb{P}(X > C) \tag{5.2}$$

where the last approximation is valid when C is large and the tail of $F(x)$, the distribution of X_t, decays fast.

A related quantity of interest is the fraction of bits lost which is just:

$$\lim_{T \to \infty} \frac{\int_0^T (X_t - C)^+ dt}{\int_0^T X_t dt} = \frac{\mathbb{E}[(X_0 - C)^+]}{\mathbb{E}[X_0]} \tag{5.3}$$

When queueing is involved the commonly considered performance measure is the buffer overflow probability that is approximated by the tail distribution of the buffer occupancy assuming the buffer is infinite given by:

$$\mathbb{P}(W_0 > B) \qquad \text{(Overflow Probability)}$$

Here B represents the finite buffer size that we would like to consider. When B is large the tail distribution is a good approximation for the overflow probability.

Thus, in both cases for measures of packet loss, we see the quantity of interest is related to the computation of tail or the complementary distribution function.

5.2 MULTIPLEXING AND EFFECTIVE BANDWIDTHS-MOTIVATION

To understand the idea of statistical multiplexing let us begin by considering a very simple scenario. This merely brings out the ideas.

Consider the case when the desired performance measure is the average delay. Consider the following $M/G/1$ model where there are N sources that are transmitting at a Poisson rate of λ packets per second. The server serves at a rate of C bits per second. The packet sizes are variable and uniformly distributed in $[0, M]$ where M represents the maximum packet size in bits.

For stability of the queue the maximum number of sources that could be supported is given by: $N\lambda \frac{M}{2} < C$ or $N < \frac{2C}{\lambda M}$.

Now consider the maximum number of sources that could be admitted if the packets were all of size M. This is given by $\frac{C}{\lambda M}$ or the fact that the packets are uniformly distributed allows the system to carry twice the number of sources if stability is the requirement.

Now let us calculate the maximum number of sources that can be admitted in the system if the bound on the mean delay is D. From the Pollaczek-Khinchine formula we obtain by noting that $\mathbb{E}[\sigma^2] = \frac{M^2}{3C^2}$ and $\mathbb{E}[\sigma] = \frac{M}{2C}$ where σ denotes the service time and equal to $\frac{\text{packet-length}}{C}$. Therefore, the mean waiting time or delay assuming a FIFO service discipline is:

$$\begin{aligned}
\mathbb{E}[W_0] &= \frac{\lambda \mathbb{E}[\sigma^2]}{2(1 - \rho)} \\
&= \frac{N\lambda \frac{M^2}{3C^2}}{2(1 - N\lambda \frac{M}{2C})}
\end{aligned} \tag{5.4}$$

The QoS requirements are $\mathbb{E}[W_0] \leq D$ or

$$N \left(\frac{\lambda M^2}{6CD} + \frac{\lambda M}{2} \right) \leq C \tag{5.5}$$

Hence, the max number of sources that can be admitted for meeting the average delay constraint is :

$$N \leq \frac{C}{\frac{\lambda M^2}{6CD} + \frac{\lambda M}{2}}$$

Clearly, as $D \to \infty$ which corresponds to an unconstrained system we see that the maximum number coincides with the number corresponding to the maximum number to guarantee stability namely $\frac{2C}{\lambda M}$.

A similar calculation assuming all of the packets are of maximum size M gives the maximum number as:

$$N \leq \frac{C}{\frac{\lambda M^2}{CD} + \lambda M}$$

Let us define the multiplexing gain denoted by MG as follows:

$$MG = \frac{\text{Max number of admitted sources with statistical variation}}{\text{Max number of sources with maxpacket size}}$$

Then we see that:

$$MG = \frac{\frac{\lambda M^2}{CD} + \lambda M}{\frac{\lambda M^2}{6CD} + \frac{\lambda M}{2}}$$

Now as $C \to \infty$ or as the link capacity grows while keeping N constant (or if C grows faster than N) we see that $MG \to 2$ which is the ratio of the number of sources assuming that the packets are uniformly distributed to the number of sources that can be admitted assuming all the packets are of maximum size if the requirement is only stability of the system.

In other words, the statistical multiplexing gain approaches the ideal gain possible as the capacity C of the server increases and thus the more the number of independent sources we can multiplex the better are the gains.

Now, equation 5.5 , can be viewed as $NA \leq C$ where the quantity

$$A = \frac{\lambda M^2}{6CD} + \frac{\lambda M}{2} \tag{5.6}$$

can be thought of as the *effective bandwidth* in analogy with loss systems. Note $\frac{\lambda M}{2}$ is just the mean load ρ and thus $\rho \leq A$ and $A \to \rho$ as $C, D \to \infty$ or the effective bandwidth is the same as the mean load when the capacity becomes large or there is no delay bound.

With this definition of A the probability that the delay bound is exceeded can be computing by considering the Erlang loss formula with $\lambda = N\lambda$, bandwidth requirement A and mean holding time 1 and taking capacity C. So the effective bandwidth can be thought as a way to map a queueing system characteristics into an equivalent loss system.

The above discussion helps to understand the notion of effective bandwidths and the fact that we can achieve gains for the same delay constraint when the capacity is large.

It can readily be seen that if we multiplex $\{n_k\}_{k=1}^M$ independent inputs, each of which generate packets at rates λ_k and which are uniformly distributed in $[0, M_k]$, then the delay bound will be met if:

$$\sum_{k=1}^M n_k A_k \leq C \tag{5.7}$$

where $A_k = \frac{\lambda_k M_k^2}{6CD} + \frac{\lambda_k M_k}{2}$ is the effective bandwidth of packets.

Thus, we define the admission region Ω as the region where admitted number of sources when multiplexed in the buffer meet their delay constraint:

$$\Omega = \left\{ \{n_k\} : \sum_{k=1}^M n_k A_k \leq C \right\} \tag{5.8}$$

In the above discussion we took the packets to be uniformly distributed but it is clear such an analysis can be carried out for any general distribution.

5.3 MULTIPLEXING FLUID INPUTS

In the previous section we considered the $M/G/1$ model. As was shown in Chapter 2, multiplexing Poisson arrival streams leads to arrivals of all classes having the same delay and thus there is no real selectivity in terms of the delay constraints that we can put. Furthermore, because of the infinite divisibility property of the Poisson process, the rate of increase of the capacity has to be faster than the rate of increase in the number of inputs in order to achieve these gains– in other words Poisson sources are not easily multiplexed.

The Poisson arrival model is not a good model for bursty behavior– Poisson processes are too *orderly*. Bursty behavior is better modeled by fluid ON-OFF models that we now consider.

In equation (2.45) which is the Pollaczek-Khinchine formula for ON-OFF sources with constant rates when they are ON, the mean is w.r.t the stationary measure. However, the delay experienced by an arriving bit from source i is the mean workload seen w.r.t. fluid Palm measure $\mathbb{P}_{A_i}$ so we need to obtain the per-class average delay which we will see is different between the classes unlike the $M/G/.$ case.

In Chapter 2, we obtained the mean delay performance observed at a FIFO queue fed with independent fluid flows. This was under the stationary measure. However, from a user perspective what is important is the mean delay of an arriving bit, and this is the mean delay calculated with respect to the fluid Palm measure. When the input flows are heterogeneous, i.e., with different (r_i, σ_i, S_i) parameters, the different classes of flows will have different mean delay even with a

FIFO scheduling policy at the queue unlike the $M/G/1$ case. Indeed, the mean delay observed at the server is a weighted average of the delay experienced by different classes of flows, which follows from the fluid conservation law and the fact that $\mathbb{P}_A = \otimes \mathbb{P}_{A_i}$. Let $\mathcal{D}$ denote the delay, in this case it is just $\frac{W_0}{c}$ where c is the server speed. Then using the fact that $A_t = \sum_{i=1}^N A_{i,t}$, we obtain

$$\mathbb{E}_A[\mathcal{D}] = \frac{1}{\lambda_A}\mathbb{E}[W_0] = \sum_{j=1}^N \frac{\lambda_j}{\lambda_A}\mathbb{E}_{A_j}[\mathcal{D}] \tag{5.9}$$

where $\mathbb{E}_{A_j}[\mathcal{D}]$ is the Palm distribution associated with the stationary increasing process $A_j(t)$ and $\lambda_A = \sum_{i=1}^M \lambda_i$.

As we will see $\mathbb{E}_{A_i}[\mathcal{D}]$ will be different for each class. Let us now obtain this.

Proposition 5.1   *In a fluid FIFO queue with N types of independent ON-OFF sources where a source of type i transmits at rate r_i when ON and the ON duration is i.i.d of length σ_i with $\mathbb{E}[\sigma_i^2] < \infty$ and the OFF durations are i.i.d exponential with $\lambda_i = \frac{r_i \mathbb{E}[\sigma_i]}{\mathbb{E}[\sigma_i]+\mathbb{E}[S_i]}$. Under the assumption that $\lambda = \sum_{i=1}^M \lambda_i < c \le r_j \ \forall j$, the mean delay experienced by the jth type of flow is given by*

$$\begin{aligned}\mathbb{E}_{A_j}[\mathcal{D}] &= \frac{\mathbb{E}[W_0]}{c} + \frac{\lambda_j \mathbb{E}[\sigma_j^2]}{2\mathbb{E}[\sigma_j]c}(r_j - \lambda_j - c + \lambda_A)(1 - p_j) \\ &= D + D_j\end{aligned} \tag{5.10}$$

where $p_i = \mathbb{P}(Source\ i\ is\ ON)$ and

$$\mathbb{E}[W_0] = \frac{1}{c - \lambda_A}\left(\sum_{i=1}^N \lambda_i \frac{\mathbb{E}_{N_i}[\sigma_i^2]}{2\mathbb{E}_{N_i}[\sigma_i]}[r_i - \lambda_i - c + \lambda_A](1 - p_i)\right) \tag{5.11}$$

Proof. Let $\{T_k^j\}$ denote the sequence of times at which the *on* periods of flow $A_j(t)$ begin and let $N_j(t) = \sum_k \mathbb{1}_{(0 < T_k^j \le t)}$ be the stationary and ergodic point process counting the number of *on-off* periods of $A_j(t)$. Let $\mathbb{P}_{N_j}$ denote the Palm measure associated with the point process $N_j(t)$. As in Chapter 3, we can obtain from the Palm inversion formulas the relationships among the average workload distributions $\mathbb{E}[W_0]$, $\mathbb{E}_{N_j}[W_0]$ and $\mathbb{E}_{A_j}[W_0]$ with respect to probability measures $\mathbb{P}$, $\mathbb{P}_{N_j}$, $\mathbb{P}_{A_j}$, respectively. We specifically have

$$\mathbb{E}[W_0] = \mathbb{E}_{N_j}[W_0] + C_1^j \quad \text{and} \quad \mathbb{E}_{A_j}[W_0] = \mathbb{E}_{N_j}[W_0] + C_2^j,$$

where

$$\begin{aligned}C_1^j &= \lambda_{N_j}\mathbb{E}_{N_j}\left[\int_0^{\sigma_{j,0}}(A(t) - ct)dt\right], \\ C_2^j &= \frac{\lambda_{N_j}}{\lambda_j}\mathbb{E}_{N_j}\left[\int_0^{\sigma_{j,0}}[A(t) - ct]A_j(dt)\right]\end{aligned}$$

with $A(t) = \sum_{j=1}^{N} A_j(t)$, $\sigma_{j,0}$ is the length of the active period and $(\lambda_{N_j})^{-1}$ the average length of one *on–off* period. Using the independence between input flows, we note that for $k \neq j$

$$\mathbb{E}_{N_j}\left[\int_0^{\sigma_{j,0}} A_k(t)A_j(dt)\right] = \mathbb{E}_{N_j}\left[\int_0^{\sigma_{j,0}} \lambda_{kt}A_j(dt)\right]$$

$$\mathbb{E}_{N_j}\left[\int_0^{\sigma_{j,0}} A_k(t)dt\right] = \mathbb{E}_{N_j}\left[\int_0^{\sigma_{j,0}} \lambda_{kt}dt\right],$$

and we obtain

$$C_2^j - C_1^j = \frac{\lambda_{N_j}}{\lambda_j}\left(F_j - E_j\right),$$

where

$$F_j = \mathbb{E}_{N_j}\left[\int_0^{\sigma_{j,0}} (A_j(t) - \lambda_j t)A_j(dt)\right] - \mathbb{E}_{N_j}\left[\int_0^{\sigma_{j,0}} (c - \lambda)t A_j(dt)\right]; \qquad (5.12)$$

and

$$E_j = \lambda_j \mathbb{E}_{N_j}\left[\int_0^{\sigma_{j,0}} (A_j(t) - \lambda_j t)dt\right] - (c - \lambda_A)\lambda_j \mathbb{E}_{N_j}\left[\int_0^{\sigma_{j,0}} t\,dt\right]. \qquad (5.13)$$

Simple computations show that

$$\begin{aligned}
C_2^j - C_1^j &= \frac{\lambda_{N_j}}{2\lambda_j}\mathbb{E}[\sigma_j^2][r - \lambda_i - c + \lambda_a](r_j - \lambda_j) \\
&= \frac{\lambda_j \mathbb{E}[\sigma_j^2]}{2\mathbb{E}[\sigma_j]}[r_j - \lambda_j - C + \lambda_A](1 - p_j)
\end{aligned} \qquad (5.14)$$

Now we use the formula for the mean of the queue workload from (2.59) to obtain

$$\mathbb{E}[W_0] = \frac{1}{c - \lambda_A}\left(\sum_{i=1}^{N} \lambda_i \frac{\mathbb{E}_{N_i}[\sigma_i^2]}{2\mathbb{E}_{N_i}[\sigma_i]}[r_i - \lambda_i - c + \lambda_A](1 - p_i)\right) \qquad (5.15)$$

Then substituting :

$$\mathbb{E}_{A_j}[W_0] = \mathbb{E}[W] + C_2^j - C_1^j \qquad (5.16)$$

and we obtain the required result. $\square$

Remark 5.2   Proposition 5.1 indicates that in a FIFO queue, the mean delay for a particular type of input consists two parts: a common part which relates to the mean delay observed at the server, and a special part which is only determined by the parameters of this type of traffic. Equation (5.10) shows traffic flows with larger burstiness or higher peak rate experience longer delay, and that flows with lower average rate also wait longer.

The above result can be used to define an admission control region that specifies the number of inputs of various types that could be supported by a link with a given capacity.

Let n_j be the number of flows of type j that can be admitted. We define the following two admission control schemes based on per-class mean delay and overall mean delay, respectively:

$$\Gamma_1 = \{\vec{n} : \mathbb{E}_{A_j}[\mathcal{D}] \le d_j, \ \text{if} \ n_j > 0, \ \text{for} \ j = 1, \cdots, N\}$$
$$\Gamma_2 = \{\vec{n} : \mathbb{E}[\mathcal{D}] \le \min_{1 \le j \le N} \{d_j\}\}.$$

For illustration, we compare the admission regions when two classes of regulated sources, specified by the parameters $(\pi_i, \rho_i, \sigma_i)$, $i = 1, 2$ (see Chapter 1), attempt to access a server. We plot the number of flows of class 1 and class 2 that can be admitted by schemes Γ_1 and Γ_2 respectively. Figures 5.4(a) and 5.4(b) illustrate two different scenarios. We see that the mean delay experienced by bits of a particular type of flow (given by Proposition 5.1) can be very different from the weighted average delay that would be obtained by considering the sum of the flows as one fluid input, i.e., the scheme Γ_2 (based on the overall mean delay) either blocks traffic that should have been admitted (Figure 5.4(a)) or admits more flows so that flows in fact can receive worse mean delay than what they demand (Figure (5.4(b))). Thus, a priori, the mean delay (under the time stationary measure) is not appropriate for defining admission control schemes.

We have thus seen the importance of working with the right probability distribution under which we compute the delay seen by an arriving bit. The main reason for the difference in the delays experienced is that even when the silent or OFF periods are exponentially distributed the arrival distribution depends on the Palm distribution associated with the particular type of flow for which the PASTA hypothesis no more holds.

Now suppose we multiplex N identical fluid sources with the assumption that $\lim_{c \to \infty} \frac{r-c}{c} = 0$ i.e., the peak rate of sources is just slightly larger than the capacity as required by the hypothesis. Suppose that $\sigma \sim Uniform(0, M)$ and we impose a delay constraint D on the mean delay. Then, noting that $(1 - p) = \frac{r-\lambda}{r}$ we can write the constraints as:

$$N[A(C, D) + \lambda C_1(D)] \le C \tag{5.17}$$

where $A(C, D) \to 0$ if $C \to \infty$ and $\lambda C_1(D) < r$ is a constant. Hence, for large capacity systems once again we have an effective bandwidth that is smaller than the peak rate.

5.4   QOS-PACKET LOSS AND EFFECTIVE BANDWIDTHS

Let us now address QoS from the perspective of packet or bit loss based on the flow level model. In modern high-speed networks such models are inadequate and a more appropriate model is one that models session arrivals as discrete events but the flow of information within a session is modelled as a continuous process whose rate is random. This leads to a fluid view of the queueing process rather than a customer-centric view.

Let us recall in such a viewpoint, the total work in the buffer or workload is now the quantity of interest. Let W_t denote the buffer content at time t. Then, for $t \ge 0$

$$W_t = W_0 + A(t) - C \int_0^t \mathbb{1}_{[W_s > 0]} ds \tag{5.18}$$

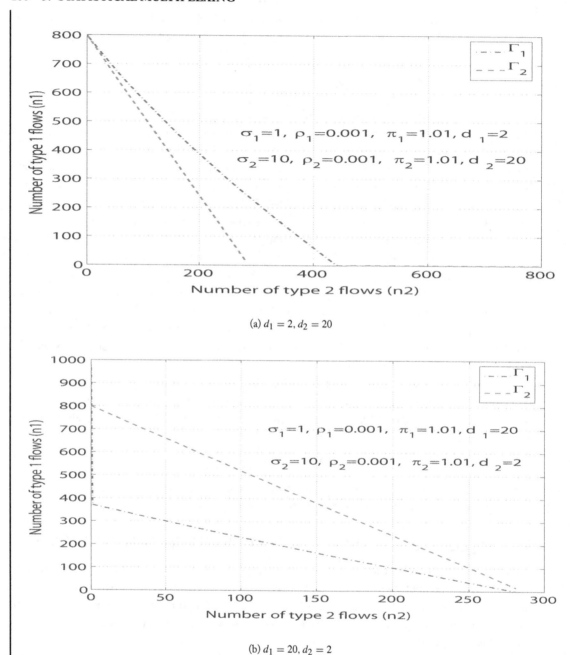

(a) $d_1 = 2, d_2 = 20$

(b) $d_1 = 20, d_2 = 2$

Figure 5.4: Admission regions Γ_1 and Γ_2: $c = 1, \sigma_1 = 1, \sigma_2 = 10, \pi_{1,2} = 1.01, \rho_{1,2} = 0.001$.

where $A(t)$ denotes the total number of bits arriving in the interval $(0, t]$ and C denotes the server rate. Note C need not be a constant or even deterministic (as in wireless links where the rate at which the link operates is random depending on the interference, channel conditions, etc.). In such a case we replace $C \int_0^t \mathbb{1}_{[W_s > 0]} ds$ by $C(t)$ where $C(t)$ denotes the total number of bits served during $(0, t]$.

Under the condition that $\rho = \mathbb{E}[A(0, 1]] < C$ from equation (2.20) we obtain:

$$W_0 = \sup_{t \geq 0} \{A(-t, 0] - Ct\}$$

When, time is discrete we can write the solution of the Lindley recursion as:

$$W_0 = \sup_{k \geq 1} \{A[-k + 1, 0] - Ck\} \tag{5.19}$$

where $A[-k + 1, 0]$ denotes the number of bits arriving in $[-k + 1, 0]$

We now focus our attention on the discrete-time model. Now to compute the overflow probability we need to compute:

$$\mathbb{P}(W_0 > B) = \mathbb{P}(\sup_{k \geq 1} \{A[-k + 1, 0] - Ck\} > B)$$

Noting the interpretation of the sup we obtain using the union bound $(\mathbb{P}(\cup A_i) \leq \sum \mathbb{P}(A_i))$ that:

$$\mathbb{P}(W_0 > B) \leq \sum_{k \geq 1} \mathbb{P}(A[-k + 1, 0] - Ck > B)$$

So we need to estimate probabilities $\mathbb{P}(A[-k + 1, 0] - Ck > B)$.

Now to estimate these probabilities we use ideas from the theory of Large deviations (assuming B and C are such that the probabilities are small so that the rhs of the union bound leads to a quantity that is non-trivial knowing that there are an infinite number of terms.

It is convenient to see this by first considering $A(-K - 1, 0] = \sum_{j=-k+1}^{0} r_j$ where r_j are i.i.d and denote the number of bits that arrive at time k. Then if we use the Chernoff bound (see Appendix)

$$\mathbb{P}(A[-k_1, 0] > Ck + B) \leq e^{-\theta B} e^{-k(\theta C - \Gamma(\theta))}$$

where $e^{\Gamma(\theta)}$ is the moment generating function of r_j. Now suppose $\frac{\Gamma(\theta)}{\theta} < C$ and calling $\alpha = C\theta - \Gamma(\theta) > 0$ we see that:

$$\mathbb{P}(W_0 > B) \leq e^{-\theta B} \sum_{k=1}^{\infty} e^{-k\alpha} = const.e^{-\theta B} \tag{5.20}$$

or we have bounded the overflow probability by a negative exponential.

One can generalize this result to the non i.i.d case by assuming that the input has stationary and ergodic increments and by defining:

$$\Gamma(\theta) = \lim_{n \to \infty} \frac{1}{n} \log \left(\mathbb{E}[e^{\theta A[-n+1, 0]}] \right)$$

and carrying out the same type of argument. Also the result given by equation 5.20 and the moment generating function now viewed as the cumulative input in an interval that is continuous can be shown when we consider the continuous-time fluid queueing model. It is only a bit more technically difficult since the union bound cannot be simply applied.

The quantity $\frac{\Gamma(\theta)}{\theta}$ is now identified as the *effective bandwidth*. Let us study some of its properties.

- If $r_j \in [0, R]$ with $\rho = E[r_j] < R$, here R is the peak rate, then

$$\rho \le \frac{\Gamma(\theta)}{\theta} < R$$

This follows from the fact that $E[e^X] \ge e^{EX}$ and so $\Gamma(\theta) \ge \theta\rho$ and $\Gamma(\theta) \le \theta R$ trivially.

So as in (5.6) we see the definition of the effective bandwidth gives a number between the mean and peak rate.

- The second property is that the effective bandwidths are additive. Let $A[-k.0] = \sum_{j=1}^{N} A_i[-k, 0]$ where A_i denote independent inputs. Then it is easy to see that $\Gamma(\theta) = \sum_{j=1}^{N} \Gamma_j(\theta)$ and so the effective bandwidths are additive.

- Finally, let us show the statistical multiplexing property, as we multiplex more and more sources then the effective bandwidth converges to the mean. For simplicity let us consider the case when there are N identical statistically independent sources accessing a buffer whose buffer size is NB and capacity is NC. Note the worst case delay given by $\frac{B}{C}$ is constant for all N. Now suppose the overflow probability bound is $\epsilon = e^{-\delta}$. Then clearly we need $\theta NB = \delta$ or $\theta = \frac{\delta}{NB}$. Now clearly $\Gamma_N(\theta) = N\Gamma(\theta)$ by independence. Hence, suppose θ is such that: $N\frac{\Gamma(\frac{\delta}{NB})}{\frac{\delta}{NB}} < NC$ then we see that as $N \to \infty$ we have

$$\lim_{N \to \infty} \frac{\Gamma(\frac{\delta}{NB})}{\frac{\delta}{NB}} \to \Gamma'(0) = \rho$$

by definition of the moment generating function. This shows that the statistical multiplexing gain that matches the max number of sources that can be admitted for stability is obtained in large systems as in the mean delay discussion above.

As above one can readily define the admission region of the number of sources that could be supported by a buffer of size B that is drained at rate C assuming that there are N types of sources, by:

$$\Omega = \left\{ \{n_k\} : \sum_{k=1}^{N} n_k \frac{\Gamma_k(\theta)}{\theta} \le C \right\} \tag{5.21}$$

where $\theta = \frac{\delta}{B}$ where $\delta = -\ln(\varepsilon)$ and ε is the bound on the overflow probability.

Note by the definition of Ω the admission region has the same form of the state-space of a multi-rate Erlang model, and so if we know the rate of arrival of the sources say λ_k for type k, assuming that they are Poisson(which can be justified and verified in practice), by computing the blocking probability for that type, we obtain the probability that the QoS constraints cannot be met. We discuss this in a bit more general and precise manner later.

Thus, one of the great advantages of the effective bandwidth is we can map the queueing model to an equivalent loss model. This is important for service providers as they can install routers with appropriate buffers and speeds to meet QoS requirements without getting into details about the queueing. Indeed, the only difficulty is that one needs the moment generating functions $\Gamma_k(\theta)$ which might be difficult to obtain in practice. So one way is to *shape* or regulate traffic flows to conform to a profile whose moment generating function is known.

The figure 5.5 indicates the type of gain in the admission region defined by the number of sources that can be admitted while meeting statistical QoS as opposed to the worst case (deterministic) results and the stability region is also indicated. Of course, if the system is large the stability region very closely approximates the admissible region as we have seen. Thus, there can be substantial gains to be had by considering statistical criteria for admission control. The drawback is that more information is required about the sources.

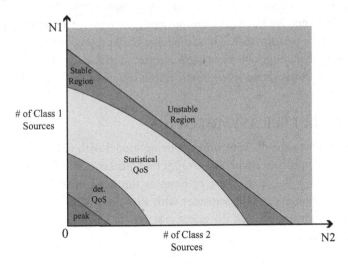

Figure 5.5: Admission regions under various criteria.

5.5   ESTIMATING BUFFER OVERFLOW PROBABILITIES

As we saw in Section 4.1 most of the performance related QoS criteria involve the calculation of the tail distribution of the queue workload or congestion process. In the previous section we saw how the overflow probability could be estimated using the Chernoff bound and this involved computing $\Gamma(\theta)$ for the input. We then saw how this function could be used to define a notion of *effective bandwidths*. In this section we consider the issue of estimating the tail distributions in more detail.

As was seen in Chapter 2, the explicit characterization of the tail distribution of the workload or congestion process except in the simple $M/M/.$ or $M/G/.$ cases is a next to impossible task. Yet this is the information we need to determine the performance and the effects of traffic characteristics. However, the often stringent performance requirements put an emphasis on the part of the tail distribution that involves small probabilities and this is fortunate because this puts us in the cadre of large deviations where sophisticated techniques exist to characterize the tail behavior. This is the issue we will now address because it leads to further understanding of how losses occur in queues and provides a very concrete framework for developing an *effective bandwidth* theory.

There are basically two different approaches to study the asymptotic behavior of the loss probability and the tail of the workload distribution that are of interest. These result from trying to address different aspects related to these overflow probabilities. The first is the *large buffer asymptotics* while the second is the *many source asymptotic*. The former is mathematically valid when the buffer size is large and there are few inputs that consume a significant portion of the bandwidth during transmission, while the latter is when a large number of independent sources access the switch or router and each transmitting source consumes a much smaller order of magnitude of the bandwidth. The second approach is more akin to the statistical multiplexing issues we have discussed in this chapter. In general, they are not equivalent but both serve to obtain better and sometimes different insights into network performance. Thus, the large buffer asymptotic is more relevant at the access of a network when there might be a few users connected to the access point while the latter are more relevant in the core or backbone where many traffic stream are multiplexed.

5.5.1   LARGE BUFFER ASYMPTOTICS

We restrict ourselves to the discrete-time queueing model with fluid inputs that are assumed to be stationary increment processes with a few other technical conditions that we will discuss.

Let $A(s, t] = \sum_{k=s+1}^{t} r_k$ where r_k denotes the amount of bits that arrive at time k. We assume $\{r_k\}$ is a stationary, ergodic sequence with $\mathbb{E}[r_k] = \rho < C$, $\forall k$ where C is the server speed. By Loynes' theorem it implies that the queue is stable. Let W_0 denote the workload in the queue at time 0. Then, we have:

$$W_0 = \sup_{t \geq 0}\{A(-t, 0] - Ct\} \tag{5.22}$$

Our aim is to calculate $\mathbb{P}(W_0 > x)$ for large values of x. Note by the SLLN, $\lim_{n \to \infty} \frac{A(-n, 0]}{n} = \rho < C$ and thus as $x \to \infty$ the tail probability goes to zero. Large deviations theory tries to capture the rate at which the tail goes to 0. However, under these general hypotheses it is a very difficult problem and thus we need to impose more hypotheses. We now state and prove the result below in the case when the input is stationary with independent increments , i.e., when the inputs $\{r_k\}$ are i.i.d.:

Proposition 5.3 *Let $A(-t, 0]$ be a stationary independent increment process with $\mathbb{E}[A(-t, 0]] < Ct$ and let $\phi(\theta) = \mathbb{E}[e^{\theta A(-1,0]}]$. Then the moment generating function of $A(-t, 0]$ is just $[\phi(\theta)]^t$. Define:*

$$I_t(x) = \sup_{\theta < \infty} \{\theta(C + \frac{x}{t}) - \ln \phi(\theta)\} \tag{5.23}$$

Then the tail distribution of the stationary workload process satisfies:

$$\lim_{x \to \infty} -\frac{\mathbb{P}(W_0 > xq)}{x} = I(q) \tag{5.24}$$

where

$$I(q) = \inf_t (t I_t(q)) \tag{5.25}$$

In other words, for large x, $\mathbb{P}(W_0 > qx) \sim Ce^{-I(q)x}$, or the tail distribution is exponentially decreasing at rate $I(q)$.

Proof. To prove this result we first note that:

$$\mathbb{P}(W_0 > xq) = \mathbb{P}\left(\sup_{t>0}(A(-t, 0] - Ct > qx)\right)$$

$$= \mathbb{P}(\cup_{t \geq 0} A(-t, 0] > Ct + qx) \leq \sum_{t>0} \mathbb{P}(A(-t, 0] > Ct + qx)$$

There are two steps: first showing the upper bound and then obtaining the lower bound and showing they are the same on the log scale. To prove the upper bound we use the Chernoff bound argument for θ such that $\phi(\theta) < \theta C$ to obtain:

$$\mathbb{P}(W_0 > qx) \leq e^{-\theta xq} \sum_{t \geq 0} e^{(t\phi(\theta)-C)} \leq e^{-\theta xq} \frac{e^{-(C-\phi(\theta))}}{1 - e^{-(C-\phi(\theta))}} = const.e^{-\theta q}$$

Hence:

$$\limsup_{x \to \infty} \frac{\log \mathbb{P}(W_0 > xq)}{x} \leq -\theta q$$

Noting by assumption $\mathbb{E}[A(-1, 0]] < C$ we see that we can take the supremum over all $\theta > 0$ on the right-hand side since it is true for all θ to obtain:

$$\limsup_{x \to \infty} \frac{\log \mathbb{P}(W_0 > xq)}{x} \leq -q \sup\{\theta > 0 : \phi(\theta) < \theta C\}$$

From the definition of $I_t(x)$ it can then be shown that the value of θ that achieves the supremum over the set $\theta C > \phi(\theta)$ is given by $I(q)$ as defined above.

The proof of the lower bound is obtained from the following observation that because of the independent increment property (assuming that x is an integer):

$$\mathbb{P}(W_0 > xq) \geq \mathbb{P}(A(-tx, 0) > Ctx + xq) \geq (\mathbb{P}(A(-t, 0] > Ct + q))^x$$

Taking logarithms we see that this implies a super-additivity property and it follows from the property of super-additive sequences that that $\lim_{x \to \infty} \frac{\log \mathbb{P}(W_0 > xq)}{x} = \sup_x \frac{\log \mathbb{P}(W_0 > xq)}{x} = I(q)$

Combining the upper and lower bounds the result follows. □

The proof in the general case of ergodic inputs is much more technically difficult and follows from a so-called *sample-path* large deviations result. We do not discuss it here suffice it to say that using this technique we can once again show that asymptotically the tail distribution has an exponential decay. An interested reader should consult the monograph of Ganesh *et al* listed at the end of the chapter for the general result.

5.5.2 MANY SOURCES ASYMPTOTICS

From the perspective of studying statistical multiplexing the *many sources asymptotic* is more appropriate and representative of describing the overflow behavior when many independent sources are multiplexed. In this case, we can obtain stronger results in that we can obtain not only the decay rates but also the constants associated with the tail distributions. Moreover, we can provide a simple proof without the need to assume that the input has independent increments.

Let us first introduce the model: Suppose there are M types of sources and there are Nm_i sources of type i accessing the buffer where N is a *scaling parameter*. The cumulative work brought in an interval of $(s, t]$ by a type i source is $A_i(s, t]$ and it is assumed that $A_i(.)$ is a stationary increment process and $\rho_i = \mathbb{E}[A_i(0, 1]]$ and $\sum_{i=1}^{M} m_i \rho_i < C$ where the total server rate is NC. Let $A^N(0, t] = \sum_{i=1}^{M} \sum_{j=1}^{Nm_i} A_{i,j}(0, t]$ where $A_{i,j}(.)$ denotes a typical source of type i. The assumption on the total average arriving work implies that the queue is stable. The scenario that is studied is depicted in Figure 5.6. This model can be viewed as an N-fold scaling while keeping the average load the same of a nominal model of buffer size B and server capacity C that is accessed by m_i, $i = 1, 2, \ldots, M$ independent sources.

By exploiting the independence and large number of sources we obtain the following asymptotics for the tail distribution of the buffer occupancy:

Proposition 5.4

Let $W_0^{(N)}$ denote the stationary workload in a work conserving queue with a fixed rate server of rate NC fed by $\{Nm_i\}_{i=1}^{M}$ stationary and mutually independent sources $\{A_i(0, t]\}_{i=1}^{M}$ that have stationary increments where sources of type i arrive at rate $\sum_{i=1}^{M} m_i \rho_i < C$ and $\rho_i = \mathbb{E}[A_i(0, 1]]$.

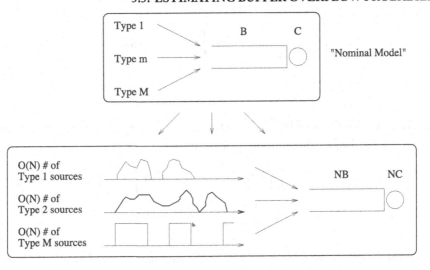

Figure 5.6: Many sources asymptotics scenario.

Define the rate function:

$$I_t(C, B) \quad = \quad \sup_{\theta}\{\theta(Ct + B) - \ln\phi_t(\theta)\}$$

$$= \quad \tau_t(Ct + B) - \ln\phi_t(\tau_t) = \tau_t(Ct + B) - \sum_{i=1}^{M} m_i \ln\phi_i(\tau_t), \qquad (5.26)$$

where:

$$\phi_t(\theta) = \prod_{i=1}^{M}(\phi_{i,t}(\theta))^{m_i} = \prod_{i=1}^{M}(\mathbb{E}[e^{\theta A_i(0,t]}])^{m_i} \qquad (5.27)$$

and τ_t is the unique solution to:

$$\frac{\phi_t'(\tau_t)}{\phi_t(\tau_t)} = Ct + B, \qquad (5.28)$$

Assume that the rate function $I_t(C, B)$ satisfies:

A1: There exists a unique t_0 such that $I_t(C, B)$ achieves its minimum at t_0.

A2: For large t $\liminf_{t\to\infty} \frac{I_t(C,B)}{\ln t} > 0$

Then, as $N \to \infty$

$$\mathbb{P}(W_0^{(N)} > NB) = \frac{e^{-NI_{t_0}(C,B)}}{\tau_{t_0}\sqrt{2\pi\sigma_{t_0}^2 N}}\left(1 + O(\frac{1}{N})\right) \qquad (5.29)$$

where:

$$\sigma_t^2 = \frac{\phi_t''(\tau_t)}{\phi_t(\tau_t)} - (Ct + u)^2 \tag{5.30}$$

Proof. From the definition of $W_0^{(N)}$ and noting from the stationary increment property of $\{A(s, t]\}$ that $A(-t, 0]$ has the sme distribution as $A(0, t]$:

$$\mathbb{P}\left\{A^N(0, t_0] > (Ct_0 + B)N\right\} \leq \mathbb{P}\left\{W_0^{(N)} > NB\right\}$$

$$\leq \sum_{t=1}^{\infty} \mathbb{P}\left\{A^N(0, t] > (Ct + B)N\right\}$$

Applying the Chernoff bound to each term in the summation, we obtain:

$$\mathbb{P}\left(A^N(0, t) > (Ct + B)N\right) \leq e^{-NI_t(C, B)}$$

Furthermore, by the uniqueness of t_0, there exists $\epsilon > 0$ such that $I_{t_0}(C, B) + \epsilon \leq I_t(C, B)$ for any $t \neq t_0$. Hence, using the Bahadur-Rao theorem (see Appendix) for large N:

$$\frac{\mathbb{P}\{A^N(0, t] > (Ct + B)N\}}{\mathbb{P}\{A^N(0, t_0] > (Ct_0 + B)N\}} \sim O(e^{-N\epsilon}) \tag{5.31}$$

Now the assumption that $\lim\inf_{t\to\infty} \frac{I_t(C,B)}{\log t} > 0$ implies that there exists $t_1 > t_0$ and $\alpha > 0$ such that $\forall t \geq t_1$ we have $I_t(C, B) > \alpha \log t > I_{t_0}(C, B)$ by definition of t_0.

Hence, for $N > \frac{1}{\alpha}$, splitting the summation into two summations from $[1, t_1]$ and the other from $[t_1 + 1, \infty)$ we obtain:

$$\sum_{t=1}^{\infty} \mathbb{P}\left\{A^N(0, t] > (Ct + B)N\right\}$$

$$\leq \mathbb{P}\left\{A^N(0, t_0] > (Ct_0 + B)N\right\} \times$$

$$(1 + t_1 e^{-N\epsilon}) + \frac{1}{N\alpha - 1}e^{(-N\alpha+1)\ln(1+t_1)}$$

Once again using Bahadur-Rao theorem we obtain the asymptotic estimate for $\mathbb{P}\{A^N(0, t_0] > N(Ct_0 + B)\}$ as:

$$\mathbb{P}\{A^N(0, t_0] > N(Ct_0 + B)\} = \frac{e^{-NI_{t_0}(C,B)}}{\tau_{t_0}\sqrt{2\pi\sigma_{t_0}^2 N}}\left(1 + O(\frac{1}{N})\right)$$

Using the above we obtain:

$$\mathbb{P}\left\{W_0^{(N)} > NB\right\} = \frac{e^{-NI_{t_0}(C,B)}}{\tau_{t_0}\sqrt{2\pi\sigma_{t_0}^2 N}}\left(1 + O\frac{1}{N}\right) \times$$

$$\left(1 + (t_1)e^{-N\epsilon} + \frac{1}{N\alpha - 1}e^{(-N\alpha+1)\ln(1+t_1)}\right)$$

Hence, for large N the second term in the brackets on the right is $1 + O(e^{-N})$ and thus the term $O(\frac{1}{N})$ dominates yielding the desired result. $\square$

Remark 5.5   The interpretation of t_0 is that t_0 is the most likely time-scale for the overflow to take place for the given input. This is dependent on the type of input.

The assumptions on the rate function are not very stringent. Assumption (A2) is valid for most commonly used traffic models including those that correspond to sessions with *long tails*. (A2) can be relaxed by requiring a growth rate $g(t)$ that goes to infinity with t provided the summation is finite.

In the next section we will see how the many sources asymptotic can be used to refine the *effective bandwidth* idea that has important practical implications and also leads to a very nice coupling with the loss level phenomena we have studied in this chapter and the loss formulae.

5.6   EFFECTIVE BANDWIDTHS REVISITED

In Section 4.4 we introduced the idea of effective bandwidths by defining the effective bandwidth as the ratio $\frac{\Gamma(\theta)}{\theta}$. Clearly, this result depends on the θ chosen and thus there are many effective bandwidths that can be associated with a single source. A natural question that one can ask is whether there is a particular choice of θ that is preferable to others? We will try to address this and other issues, in particular give a nice interpretation for effective bandwidths related to the admission region Ω that we defined earlier.

Let us consider a model where sessions access a link and the duration of sessions is now random and of finite mean. The scenario we consider is that of the many sources asymptotic except that sessions of type i arrive as a Poisson process with rate $N\lambda_i$. An arriving session of type i transmits according to the process $A_i(0, t]$ in an interval of length $(0, t]$. Let us assume the mean duration of transmission is 1 unit and the average load brought by a session of type i is $\sigma_i = \mathbb{E}[A_i(0, 1]]$ and is such that $\sum_{i=1}^{M}\lambda_i\sigma_i < C$.

Let **m** denote the vector of the number of on-going connections. Define

$$\Omega_\varepsilon = \{\{m_i\}_{i=1}^{M} : \mathbb{P}(W_0 > NB) \le \varepsilon, A(0, t] = \sum_{i=1}^{M}\sum_{j=1}^{m_i}A_{i,j}(0, t]\}, \tag{5.32}$$

i.e., Ω_ε corresponds to the configuration of active sessions accessing the buffer that meet the constraint on overflow probability- the admissible number of sessions that can access the buffer simultaneously. Let $conv(\Omega_\varepsilon)$ denote the convex hull of Ω_ε. Then one can show that the resulting system in equilibrium has the following product-form solution:

$$\pi(\mathbf{m}) = \frac{1}{G} \prod_{i=1}^{M} \frac{(N\lambda_i)^{m_i}}{m_i!} \tag{5.33}$$

where G is the normalizing factor obtained by summing the numerator over $conv(\Omega_\varepsilon)$.

Define $\mathbf{m}^* = (m_1^*, m_2^*, \ldots m_M^*)$ as arg max $\pi(\mathbf{m})$ such that $\mathbb{P}(W_0 > NB) \approx \varepsilon$. This has the interpretation of the *most likely overflow configuration*. Of course the requirement that m_i be an integer means that we choose configurations such that they lie close to the boundary of the convex hull of the admissible region. Then we can show that $\frac{\pi(\mathbf{m}^*)}{\pi(\mathbf{m})} = O(e^{-N})$ for any other configuration on the boundary, implying that the most likely configuration dominates. In other words, in equilibrium we are most likely to see the system in $\mathbf{m}^*$ when an overflow takes place.

We then define the effective bandwidth $A_i(N)$ for class i as the gradient in direction i of the tangent hyperplane at $\mathbf{m}^*$.

Then we can define an admissible region as:

$$\hat{\Omega}_\varepsilon = \{\mathbf{m} : \sum_{i=1}^{M} A_i(N)m_i \le C + \frac{B}{t_0} = C^*\} \tag{5.34}$$

C^* is now the effective service rate. Essentially the effect of the buffer is to provide a slight increase of capacity and $\frac{B}{t_0}$ represents the amount of additional rate that can be supported before information is lost. Pictorially the procedure is shown in Figure 5.7.

Let $A_j(N)$ denote the effective bandwidth obtained via this procedure. Then it can be shown that $\lim_{N\to\infty} A_j(N) = \lambda_j \mathbb{E}[A_j(0,1]]$ or in other words as the system becomes larger the effective bandwidth of a source can be approximated by its mean rate of arriving load. It can also be shown that $A_i(N)$ depends only on $\phi_i(\theta)$, $\{\lambda_i\}_{i=1}^{M}$, B and C.

The remarkable property is that if we compute the probability of a connection being blocked, i.e., probability that an arriving connection cannot be accommodated in Ω_ε defined by the overflow probability, then the corresponding blocking probabilities for the multi-rate Erlang loss model (given in Proposition 3.8) using the $A_i's$ corresponding to the effective bandwidths and capacity C^* are almost identical (within a factor $O(\frac{1}{N})$) implying there is a consistent mapping from the overflow probability to the blocking probability for multi-rate Erlang loss models via the effective bandwidths as defined. In other words, the introduction of the effective bandwidth allows us to ignore the queueing details and use the multi-rate Erlang formula for allocating resources. The details can be found in the article Likhanov, Mazumdar and Theberge given below.

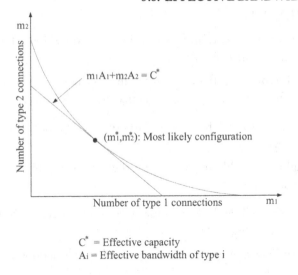

C^* = Effective capacity

A_i = Effective bandwidth of type i

Figure 5.7: Effective bandwidths as hyperplane approximation to acceptance boundary.

CONCLUDING REMARKS:

The aim of this chapter has been to show the interplay between the queueing models of Chapter 2 and the QoS issues in networks. In particular, the multiplexing of a large number of traffic sources leads to *economies of scale* in that if we allow for some loss or delay tolerance then we can accommodate many more sources than can be done if we reserve bandwidth based on the worst case or peak behavior. These ideas can readily be extended to queueing models with variable rate servers such as those that arise in wireless networks due to the noisy nature of the channel that leads to variable bit rates. The applications considered in this chapter are some of the key ideas that are used for admission control. Essentially the effective bandwidth captures the bandwidth requirements of a source to meet its QoS requirements in terms of loss or delay. In other words, the switch or router must guarantee that at least the bandwidth specified by a source's effective bandwidth must be reserved throughout the connection duration.

This concludes our discussion of statistical multiplexing. It is hoped that the modeling and analysis ideas that have been developed will lead the reader to explore the topics further. The references cited offer many more references and examples than have been considered here.

NOTES AND PROBING FURTHER

As mentioned in the introduction, the notion of *effective bandwidths* has been one of the key ideas that emerged during the study of admission control for ATM networks. The original idea is due to Hui for bufferless (loss) systems and has been expanded and generalized for queues by Kelly. These ideas have only just begun to percolate into text books but not generally widespread.

BOOK REFERENCES

F.P. Kelly, *Notes on effective bandwidths*, in Stochastic Networks, Kelly, F.P., Zachary, S., and Zeidins, I., editors, Oxford University Press, 1996

Most of he ideas on the buffer asymptotics in this chapter can be found in these notes. It is an extremely useful reference for results on effective bandwidths for different types of sources.

A. Kumar, D. Manjunath, and J. Kuri: *Communication Networking: An analytical approach*, Morgan-Kaufman (Elsevier), 2004.

This is a highly accessible treatment and there are many nice problems related to effective bandwidth calculations.

A. Ganesh, N. O'Connell, and D. Wischik; *Big Queues*, Lecture Notes in Mathematics 1838, Springer-Verlag, Berlin, 2004.

This is an excellent monograph on the use of large deviations techniques to study overflow problems in queues.

N. B. Likhanov, R. R. Mazumdar, and F. Theberge, *Providing QoS in Large Networks: Statistical Multiplexing and Admission Control*, in *Analysis, Control and Optimization of Complex Dynamic Systems*, E. Boukas and R. Malhame , eds., Springer 2005, pp 137-1169.

This paper explores the effective bandwidths in the context of large systems and shows how one obtains a consistent mapping from the queueing level to the loss network model as discussed in the last section. In particular, the multiplexing gain as viewed from the perspective of the admission region for large networks is studied in detail.

JOURNAL ARTICLES

J.Y. Hui, *Resource allocation in broadband networks*, IEEE Journal of Selected Areas in Communications, 1989, 6:1598–1608.

This is the original paper where the idea of effective bandwidths first appeared. Although done for unbuffered systems the concept is the same.

F.P. Kelly, *Effective Bandwidths at Multiclass Queues*, Queueing Systems 9 (1991), pp. 5-16.

This paper generalized Hui's idea to queues. It is a very seminal paper.

G. Kesidis, J. Walrand, and C.S. Chang,*Effective bandwidths for multiclass Markov fluids and other ATM sources*, IEEE/ACM Transactions on Networking (TON), 1993, Vol 1, 424-428

D. D. Botvich and N. G. Duffield, *Large deviations, the shape of the loss curve, and economies of scale in large multiplexers*. Queueing Systems Theory Appl. 20 (1995), no. 3-4, 293–320.

This paper first presented a comprehensive analysis of statistical multiplexing showing that so-called economies of scale or multiplexing gains can be achieved by multiplexing many independent bursty sources. They identify multiplexing gains of various types of bursty inputs through determining their loss curves.

A. El Walid and D. Mitra, *Effective bandwidth of general Markovian traffic sources and admission control of high speed networks*,IEEE/ACM Transactions on Networking (TON), Volume 1 , Issue 3 (June 1993),pp. : 329 - 343

This paper presents the applications of effective bandwidths to admission control.

APPENDIX A

Review of Probability and Markov Chains

In this appendix we collect some results from probability and stochastic processes that are used in this monograph. Only the results that have been directly used are recalled here.

A.1 LIMIT THEOREMS, LAWS OF LARGE NUMBERS

These set of results relate to the so-called ergodic theorems which relate the limits of empirical averages of sums of identically distributed random variables or integrals of stationary stochastic processes to their means. The means are computed under the probability distribution under which the process is stationary (or referred to as the invariant probability distribution). So for example if the underlying probability space is $(\Omega, \mathcal{F}, \mathbb{P})$ and a process is stationary with respect to time shifts θ_t i.e., $\mathbb{P}(X_0 \circ \theta_t(\omega) \in A) = \mathbb{P}(X_0(\omega) \in A)$ for every $A \in \mathcal{F}$ then the Strong Law of Large Numbers (SLLN) states that $\lim_{T \to \infty} \frac{1}{T} \int_0^T X_s ds = \mathbb{E}[X_0] = \int_\Omega X_0(\omega) dP(\omega)$. Of course, it is important to know when the SLLN holds. When $\mathbb{P}_N$ is the underlying distribution or the underlying probability is the Palm probability associated with some stationary point process (also compatible w.r.t. θ_t meaning that N_t has stationary increments) then we have the computation of the mean under the Palm probability, and so on.

Below are the statements of the main results in the discrete and continuous time case.

Proposition A.1 *(Strong Law of Large Numbers (SLLN))*

a. Let $\{X_n\}$ be a second order sequence with mean $\mathbb{E}[X_k] = m_k$ and covariance $R(j, k)$. Suppose $R(j, k)$ satisfies for $|j - k|$ large:

$$|R(j, k)| \leq Cg(|j - k|)$$

such that

$$\frac{1}{n} \sum_{|k| \leq n-1} g(k)\left(1 - \frac{|k|}{n}\right) \leq \frac{C}{n^\alpha} \; ; \alpha > 0$$

Then,

$$\frac{1}{n} \sum_{k=0}^{n-1} (X_k - m_k) \overset{a.s}{\to} 0$$

b. Let $\{X_t\}$ be a second-order continuous process with $\mathbb{E}[X_t] = m_t$ and covariance $R(t, s)$. If for large $|t - s|$

$$|R(t, s)| \le Cg(|t - s|)$$

with

$$\frac{1}{T} \int_0^T g(t)(1 - \frac{|t|}{T})dt \le \frac{c}{T^\gamma} \; ; \gamma > 0$$

Then:

$$\frac{1}{T} \int_0^T (X_s - m_s)ds \overset{a.s.}{\to} 0$$

Remark A.2   In the case of discrete sequences, the conditions are trivially verified if $\{X_n\}$ are i.i.d sequences with $\mathbb{E}[X_n^2] < \infty$. Actually in the i.i.d. case we only need $\mathbb{E}|X_n| < \infty$.

In the continuous case (as we shall note later) the conditions are true when $\{X_t\}$ is positive-recurrent Markov process. As archetype of this is when X_t has stationary and independent increments, i.e., it is a Levy process.

The second set of limit theorems that are important are the Central Limit Theorem (CLT) and its stronger version called a local limit CLT that holds for densities.

Proposition A.3   *(Central Limit Theorem)*
Let $\{X_i\}_{i=1}^n$ be a sequence of i.i.d. r.v's with $\mathbb{E}[x_i^2] < \infty$. Define $S_n = \sum_{i=1}^n X_i$. Then as $n \to \infty$:

$$\mathbb{P}\left(\frac{S_n - nm}{\sigma\sqrt{n}} \le x\right) \to \Phi(x)$$

where $\mathbb{E}[X_1] = m$ and $var(X_1) = \sigma^2$ and $\Phi(x)$ denotes the standard normal distribution given by:

$$\Phi(x) = \frac{1}{\sqrt{2\pi}} \int_{-\infty}^x e^{-\frac{y^2}{2}} dy$$

Proposition A.4   *(Local limit theorem)*
Let $\{X_i\}_{i=1}^n$ be a collection of i.i.d. mean 0 and variance 1 random variables. Suppose that their common characteristic function $\phi(.)$ satisfies:

$$\int_{-\infty}^{\infty} |\phi(t)|^r dt < \infty$$

for some integer $r \geq 3$. Then a density $p_n(x)$ exists for the normalized sum $\frac{S_n}{\sqrt{n}}$ for $n \geq r$ and furthermore:

$$p_n(x) = \frac{1}{\sqrt{2\pi}} e^{-\frac{x^2}{2}} \left(1 + O(\frac{1}{n^{\frac{r-1}{2}}})\right)$$

Remark A.5 Actually the error term in the local limit theorem can be made more precise given that moments of order $r \geq 2$ exist under the assumption on the characteristic function.

Finally, we recall some simple ergodic theorems due to Birkoff that are weaker than the SLLN.

Definition A.6 A set A is said to be invariant for a process $\{X_t\}$ if $\{X_s \in A; s \in \Re\}$ implies that $\{X_{t+s} = X_s \circ \theta_t \in A; s \in \Re\}$ for all $t \in \Re$. In other words, the shifted version of the process remains in A.

An example of an invariant set is :

$$A = \{X. : \lim_{T \to \infty} \frac{1}{T} \int_0^T f(X_{s+t})dt = 0\}$$

An example of a set which is not invariant is the following:

$$A = \{X. : X_s \in (a, b) \; for \; some \; s \in [0, T]\}$$

Definition A.7
A stochastic process $\{X_t\}$ is said to be ergodic if every invariant set A of realizations is such that $\mathbb{P}(A) = 0$ or $\mathbb{P}(A) = 1$.

Remark: An ergodic process need not be stationary. For example a deterministic process can be ergodic but is not stationary.
If the process is strictly stationary the following theorem can be stated.

Proposition A.8 *Let $\{X_t\}$ be a strictly stationary process and $f(.)$ be a measurable function such that $\mathbb{E}[|f(X_t)|] < \infty$. Then*

a) The following limit :

$$\lim_{T \to \infty} \frac{1}{T} \int_0^T f(X_s)ds$$

exists almost surely and is a random variable whose mean is $\mathbb{E}[f(X_0)]$.

b) If in addition the process is ergodic then:

$$\frac{1}{T} \int_0^T f(X_s)ds \overset{a.s.}{\to} \mathbb{E}[f(X_0)]$$

These are the main results that have been used in the notes. Proving ergodicity is often a challenging problem. Stationarity on the other hand is easier to show.

A.2   MARKOV CHAINS AND REVERSIBILITY

Markov chains (in both discrete-time and continuous-time) are the natural building blocks for modeling stochastic systems because the evolution dynamics driven by independent sequences or independent increment processes usually result in Markov chains and processes if there is no infinite dependence on the past. Markov chains (MC) is a term reserved for the case when the process takes a finite or countable set of values called the state space , usually denoted by E. Since MC play such an important role in queueing theory we list some basic results.

A.2.1   DISCRETE-TIME MARKOV CHAINS

Definition A.9   Let $\{X_n\}$ be a discrete-line stochastic process which takes its values in a space E. Let $A \subset E$. If

$$\mathbb{P}\{X_n \in A | X_o, X_1, \ldots X_{n-1}\} = P\{X_n \in A \mid X_n\}$$

More generally

$$\mathbb{P}\{x_n \in A \mid \mathcal{F}_n^x\} = \mathbb{P}\{x_n \in A \mid x_n\}$$

where $\mathcal{F}_n = \sigma\{X_u, u \le n\}$ the sigma-field of all events generated by the process $\{X_k\}$ up to n. then $\{X_n\}$ is said to be a discrete-time Markov process.

When

$$E = \{0, 1, \ldots, \}$$

i.e., a countable set then $\{X_n\}$ is said to be a Markov chain.

The conditional probability

$$P\{X_{k+1} = j | X_k = i\} = P_{ij}(k)$$

is called the transition probability of the Markov chain and $P = \{p_{i,j}(k)\}_{(i,j) \in E \times E}$ is called the transition probability matrix.

If the transition probability does not depend on k then the chain is said to be homogeneous.

Let $\pi_i(n) = \mathbb{P}(X_n = i)$ denote the distribution of the chain at time n and let $E = \{0, 1, \ldots, \}$. Let $\pi(n) = (\pi_0(n), \pi_1(n), \ldots, \pi_k(n), \ldots)$ denote the row vector.

We now quickly recall some of the principal results about discrete-time MC. We assume that it is homogeneous.

An important property of MC is a so-called strong Markov property that states that a MC when viewed at appropriate random times retains the Markov property as well as its transition structure. These random times are called stopping times or Markov times if have the following property:

Definition A.10   A random time τ is said to be a Markov or stopping time if the event $\{\tau = n\}$ can be completely determined by knowing $\{X_0, X_1, \ldots X_n\}$, for example

$$P\{\tau = n \,|X_m, \, m \geq 0\} \;=\; P\{\tau = n|X_m, \, m \leq n\}$$

So in particular : $\tau = f(X_u, u \leq n)$ for some measurable function $f(.)$.

More generally if $\mathcal{F}_n^X$ denotes a filtration generated by a sequence $\{X_k\}^1$ then a r.v. $\tau(\omega)$ is said to be a stopping time if the event

$$\{\tau = n\} \in \mathcal{F}_n$$

So if τ is independent of $\{X_n\}$ then it is trivially a stopping time as it is contained in $\mathcal{F}$
The strong Markov property is the following :

Proposition A.11   *(Strong Markov Property)*

Let $\{X_n\}$ be a homogeneous M.C. on $(E, \, P)$. Let τ be a stopping time relative to $\{X_k\}$.

1. *The processes X_n before and after τ are independent given X_τ.*

2. *$P\{X_{\tau+k+1} = j \mid X_{\tau+k} = i\} = P_{ij}$*
   (i.e. the process after τ is an M.C. with the same transition probability P).

The importance of stopping times is also in the context of random sums of random variables, often called Wald's identity .

Proposition A.12   *Let $\{X_n\}$ be a sequence of i.i.d. random variables with $\mathbb{E}|X_0| < \infty$ and $\tau(\omega)$ be a stopping time relative to $\mathcal{F}_n^X$ where $\mathcal{F}_n^X = \sigma\{X_u, u \leq n\}$, i.e., the history or filtration generated by X up to n. Then:*

$$\mathbb{E}[\sum_{k=1}^{\tau} X_k] = \mathbb{E}[\tau]\mathbb{E}[X_0] \tag{A.1}$$

---

[1]An increasing family of σ fields indexed by n

If in addition $\mathbb{E}[X_0^2] < \infty$ then:

$$\mathbb{E}[\sum_{k=1}^{\tau} X_k - \tau\mathbb{E}[X_0]]^2 = var(X_0)\mathbb{E}[\tau] \tag{A.2}$$

Remark A.13   The i.i.d. hypothesis can be slightly relaxed if we assume that $\{X_k\}$ is a martingale difference sequence.

Let N_i denote the number of visits of $\{X_n\}$ to a given state $i \in E$. Then $N_i = \sum_n \mathbb{1}_{[X_n=i]}$.

Definition A.14   A state i is said to be recurrent if $N_i = \infty$ a.s.. Let T_i be the first return to i defined as:

$$T_i = \inf\{N > 0 : X_n = i | X_0 = i\}$$

If $\mathbb{E}_i[T_i] < \infty$ then the state is said to be positive recurrent while if $\mathbb{E}_i[T_i] = \infty$ the i is said to be null recurrent. A state that is not recurrent is said to be transient.

A MC is said to be irreducible if all states in E can communicate with each other and one can show that the property of recurrence (positive or null) is inherited by all states in E.

We then have the main result which is referred to as an ergodicity result.

Proposition A.15   *Let $\{X_n\}$ be a homogeneous positive recurrent MC. Then:*

$$\lim_{n\to\infty} \pi(n) \to \pi$$

where:

$$\pi = \pi P \leftrightarrow \tag{A.3}$$
$$\pi_j = \sum_{i\in E}\pi_i P_{i,j} \ \ j \in E \tag{A.4}$$

Moreover, the chain is ergodic.

This completes our quick view of discrete-time MC.

A.2.2 MARKOV CHAIN CONTINUOUS-TIME

Let us begin by a quick overview of continuous-time Markov chains (CTMC). We are only interested in non-negative chains.

Definition A.16 A stochastic process $\{X_t\}$ is said to be a continuous-time MC with values in $\{0, 1, 2, \cdots, \}$ if for $t > s$:

$$\mathbb{P}\{X_t \in B | X_u, u \leq s\} = \mathbb{P}\{X_t \in B | X_s\}$$

Let $P(t, s)$ denote the transition probability where:

$$P(t, s) = \{P_{i,j}(t, s)\}$$

and

$$P_{i,j}(t, s) = \mathbb{P}(X_t = j | X_s = i)$$

The CTMC is said to be homogeneous if $P(t, s) = P(t - s)$.

Now let s be fixed (say 0) and let us call $P(t, 0) = P_t$. Then it can be shown that:

$$P_0 = I, \quad \frac{dP_t}{dt} = QP_t = P_t Q$$

where $Q = \{q_{i,j}\}$ is called the infinitesimal generator of X_t and has the interpretation:

$$\mathbb{P}(X_{t+dt} = j | X_t = i) = q_{i,j} dt + o(dt)$$

By definition of Q we have $\sum_j q_{i,j} = 0$ or $q_{i,i} = -\sum_{j \neq i} q_{i,j}$

One interpretation of $-q_{i,i}$ is that $\frac{1}{-q_{i,i}}$ is the average time of the MC in state i (so $-q_{i,i}$ is the rate at which it leaves state i).

A MC is said to be in equilibrium if it is stationary. Let Π denote the stationary distribution where: $\Pi_j = \mathbb{P}\{X_0 = j\} > 0$ *and* $\sum_j \Pi_j = 1$. In particular:

$$\Pi Q = 0 \ \ i.e., \ \ \sum_i \Pi_i q_{i,j} = 0$$

If $\sup_i -q_{i,i} < \infty$ the process is said to be stable. If the process is irreducible, and positive recurrent then it is ergodic. Then the following are true:

$$\mathbb{P}(X_t = j) \to \Pi_j \qquad as \ t \to \infty \ (Limiting \ distribution)$$

$$\lim_{T \to \infty} \frac{1}{T} \int_0^T \mathbb{1}_{[X_t = j]} dt \ = \ \Pi_j \ (Ergodicity)$$

The study of conditions for a MC to be positive recurrent is one of the primary foci in the study of MCs. We will not go into this direction suffice it to say that we will be primarily interested in the stationary setting.

If a process is Markov in forward time then it is Markov in reverse time. Here is a result that relates the generator of a Markov process in forward time to the reverse time process.

Proposition A.17   *Suppose $\{X_t\}$ is a stationary Markov process with equilibrium distribution Π. Then X_{-t} (the reversed process) is stationary Markov with equilibrium distribution Π and infinitesimal generator R where:*

$$r_{i,j} = \frac{\Pi_j}{\Pi_i} q_{j,i} \tag{A.5}$$

An even stronger notion is that of a reversible Markov process :

Definition A.18   A Markov process is said to be reversible if the infinitesimal generators of the forward and reversed process are the same.

The following condition, called detailed balance (or often the balance equation), is a necessary and sufficient condition for a MC in equilibrium to be reversible.

Proposition A.19   *$\{X_t\}$ is reversible if and only if:*

$$\Pi_i q_{i,j} = \Pi_j q_{j,i} \tag{A.6}$$

How does one verify that a Markov chain is reversible without computing its stationary distribution first? A condition that only depends on the structure of the transition probabilities was given by Kolmogorov and called Kolmogorov's loop condition that we state below.

Proposition A.20   *Consider an aperiodic irreducible Markov chain with transition rates q_{ij} and stationary distribution π. Then the chain is reversible if and only if the Kolmogorov's loop criterion holds: for all $i_1, i_2, ..., i_m \in E$ and all $m \geq 3$,*

$$q_{i_1 i_2} q_{i_2 i_3} \cdots q_{i_{m-1} i_m} = q_{i_1 i_m} q_{i_m i_{m-1}} \cdots q_{i_2 i_1} \tag{A.7}$$

Remark A.21   It can be shown that in order for a Markov chain to be reversible the transition rates have the form:

$$\frac{q_{ij}}{q_{ji}} = \frac{g(j)}{g(i)}$$

for some function $g(.)$.

There is a very interesting property associated with a reversible MC that plays an important role in developing formulae for loss systems that is stated below.

Proposition A.22 *Let $\{X_t\}$ be a reversible MC in equilibrium. et A be any closed subset of its state-space. Then suppose we restrict the process to A (by setting all transitions from A to $A^c = S - A$ to be 0). Let $\bar{\Pi}$ denote the stationary distribution of the restricted process. Then:*

$$\bar{\Pi}_k = \frac{\Pi_k}{\sum_{j \in A} \Pi_j}, \quad k \in A$$

A very simple but rather powerful class of processes that occur in queueing models are the so called birth-death models which are Markov chains whose infinitesimal generator has a diagonal structure i.e., only $q_{i,i+1}$, $q_{i,i}$, and $q_{i,i-1}$ are non-zero.

Another important property of Markov chains it the property of uniformization . It is a mathematical device by which we can convert a CTMC to a DTMC such that the stationary distributions coincide.

Let $\{X_t\}$ be a aperiodic, irreducible, homogeneous CTMC on E. Suppose $\sup_i q_i = q < \infty$ and define: Define the matrix:

$$R = I + \frac{1}{q} Q$$

By construction the row sums of R are all 1, all elements are take values in $(0, 1)$ and therefore R is a stochastic matrix that can be associated with a DTMC, say Y_n.

Then the following result holds termed uniformization of the CTMC.

Proposition A.23 *Let $\{X_t\}$ be an aperiodic, homogeneous CTMC with generator Q and $\sup_i -q_i = q < \infty$. Let $\{Y_n\}$ be a DTMC on the same state space with transition probability matrix defined by $R = I = \frac{Q}{q}$. Then the stationary distributions of X_t and Y_n are equal, i.e.;*

$$\mathbb{P}(X_0 = i) = \pi_X(i) = \mathbb{P}(Y_0 = i) = \pi_Y(i), \quad i \in E$$

We conclude our discussion of Markov processes with a discussion of martingales and stochastic intensities associated with CTMCs .

Let $\{X_t\}$ be a homogeneous CTMC with infinitesimal generator Q defined on E. Then let $\mathcal{F}_t$ denote the right-continuous filtration generated by X_s, $s \le t$. Define the semi-group $P_{t-s} f(X_s) = \mathbb{E}[f(X_t)|\mathcal{F}_s]$ where $f(.)$ is a bounded continuous function.

Proposition A.24 *Then for any $T > 0$ and $t \le T$, the process:*

$$M_t = P_T f(X_0) - P_{T-t} f(X_t), \quad t \ge 0 \tag{A.8}$$

is a $\mathcal{F}_t$- martingale.

   If $\sup_i q_i < \infty$, *then Q is the generator of the semi-group and :*

$$m_t = f(X_t) - f(X_0) - \int_0^t Qf(X_s)ds$$

is a $\mathcal{F}_t$- martingale.

   In particular, if we consider the point process $N_{ij}(t) = \mathbb{1}_{[X_{t-}=i,X_t=j]}$ that counts the transitions from i to j then: $N_{ij}(t) - q_{ij} \int_0^t \mathbb{1}_{[X_s=i]}ds$ is a $\mathcal{F}_t$ martingale.

   With this we end our brief (and rather incomplete) background material in Markov processes.

A.3   EXPONENTIAL TWISTING AND LARGE DEVIATIONS

We conclude the appendix by recalling some results that are used in the notes. These have to do with the analysis of large loss systems and asymptotics of queues.

   The first important notion is that of a measure change or absolute continuity of probability measures.

Definition A.25   Let P and Q be two probability measures defined on $(\Omega, \mathcal{F}, \mathbb{P})$. Then P is said to be absolutely continuous with respect to Q, denoted by $P \ll Q$ if for any $A \in \mathcal{F}$, $Q(A) \leq \varepsilon \Rightarrow$ $P(A) \leq \delta$ for ε, δ small.

   If $P \ll Q$ and $Q \ll P$ then P and Q are said to be equivalent.

   We then have the following result:

Proposition A.26   *Let P and Q be two probability measures defined on $(\Omega, \mathcal{F}, \mathbb{P})$ and $P \ll Q$ then: there exists a non-negative r.v $\rho(\omega)$ called the Radon–Nikodym derivative and denoted:*

$$\frac{dP}{dQ}(\omega) = \rho(\omega) \tag{A.9}$$

and for any r.v. $X(\omega)$ such that $\int |X(\omega)|dP(\omega) < \infty$ we have:

$$\mathbb{E}_P[X] = \int X(\omega)dP(\omega) = \int X(\omega)\rho(\omega)dQ(\omega)$$

Remark A.27   The absolute continuity is assured if densities exist.

   Three important results that we use in book are related to exponential twisting, the Chernoff bound, and the theorem of Bahadur-Rao.

Lemma A.28   *(Exponential twisting or centering)*

Let $F(x)$ denote a probability distribution on $\Re$ with moment generating function $M(h)$. Define a new probability distribution

$$dF_a(x) = \frac{e^{h_a x}}{M(h_a)} dF(x)$$

where h_a is the point where the supremum of

$$I(a) = \sup_h \{ha - \log M(h))\} \qquad (A.10)$$

is achieved.

Then $F_a(x)$ defines a probability distribution such that the mean of the r.v. X under F_a is a, i.e., $\int x \, dF_a(x) = a$ and $var_a(X) = \frac{M''(h_a)}{M(h_a)} - a^2$.

Remark A.29   The term $\frac{e^{h_a x}}{M(h_a)}$ is the Radon-Nikodym derivative of F_a w.r.t. F. It allows to *center* the original distribution around a no matter what the original mean is.

We now state Cramer's theorem in which the right-hand side is called the Chernoff bound.

Proposition A.30   *Let X be a r.v. whose moment generating function , M(h), is defined for $h < \infty$. Then :*

i) If $a > m$, $\Pr(X \geq a) \leq e^{-I(a)}$

ii If $a < m$, $\Pr(X \leq a) \leq e^{-I(a)}$

The function $I(a)$ is called the rate function associated with the distribution $F(.)$.

We conclude with stating the Bahadur-Rao theorem which combines the local limit theorem with the exponential centering idea and a local limit version due to Petrov.

Proposition A.31
Let $S_N = \sum_{j=1}^{N} X_j$ where $\{X_i\}$ are i.i.d. r.v.'s with moment generating function $\phi(h)$. Then as $N \to \infty$ uniformly for any $u > 0$,

$$\mathbb{P}(S_N \in (Nu, Nu + du)) = \frac{e^{-NI(u)}}{\sqrt{2\pi\sigma^2 N}} du \left(1 + O(\frac{1}{N})\right), \qquad (A.11)$$

and for $u > \mathbb{E}[X_j]$

$$\mathbb{P}[S_N > Nu] = \frac{e^{-NI(u)}}{\tau_u \sqrt{2\pi\sigma_u^2)N}} \left(1 + O(\frac{1}{N})\right) \qquad (A.12)$$

where

$$I(u) = u\tau_u - \ln \phi(\tau_u) \qquad (A.13)$$

τ_u *is the unique solution to*

$$\frac{\phi'(\tau_u)}{\phi(\tau_u)} = u \qquad\qquad (A.14)$$

and

$$\sigma_u^2 = \frac{\phi''(\tau_u)}{\phi(\tau_u)} - u^2 \qquad\qquad (A.15)$$

Remark A.32   The first result is due to Petrov and the second result is due to Bahadur-Rao. The first is the approximation by a density while the second is on the distributions.

BIBLIOGRAPHY

G. R. Grimmett and D. R. Stirzaker, *Probability and Random Processes*, Oxford Science Publ, 1998

H. Cramér, and M.R.Leadbetter, *Stationary and related stochastic processes: Sample function properties and their applications*, Dover Publications Inc., N.Y., 2004 (Reprint of the 1967 original).

J. R. Norris, *Markov Chains*, Cambridge University Press, 1997

V. V. Petrov, *Sums of independent random variables*, Translated by A. A. Brown, Springer-Verlag, New York-Heidelberg, 1975

R. R. Bahadur and R. Ranga Rao, *On deviations of the sample mean*, Ann. Math. Statist., 31, 1960, pp. 1015-1027.

These books and the last paper contain all the results as do many other excellent texts on probability and stochastic processes.

Bibliography

This bibliography collects all the references used in the monograph. They are separated into book references and key papers that have appeared in journals. The bibliography is not exhaustive.

BOOKS AND EDITED VOLUMES

S. Asmussen, *Applied probability and queues*, Applications of Mathematics 51, Springer-Verlag, N.Y., 2003

F. Baccelli and P. Brémaud; *Elements of Queueing Theory: Palm-Martingale Calculus and Stochastic Recurrences*, 2nd Ed., Springer-Verlag, N. Y., 2005

M. Bramson, *Stability of Queueing Networks*, Lecture Notes in Mathematics, Vol 1950, Springer, Berlin, 2008

P. Brémaud, *Point processes and queues. Martingale dynamics*. Springer Series in Statistics. Springer-Verlag, New York-Berlin, 1981.

C-S. Chang; *Performance Guarantees in Communication Networks*, Springer-Verlag, London, 2000.

H.C. Chen and D.D. Yao, *Fundamentals of Queueing Networks: Performance, Asymptotics, and Optimization*, Applications of Mathematics 46, Springer-Verlag, NY, 2001.

H. Cramér, and M.R.Leadbetter, *Stationary and related stochastic processes: Sample function properties and their applications*, Dover Publications Inc., N.Y., 2004 (Reprint of the 1967 original).

A. Ganesh, N. O'Connell, and D. Wischik; *Big Queues*, Lecture Notes in Mathematics 1838, Springer-Verlag, Berlin, 2004.

A. Girard,*Routing and Dimensioning in Circuit-Switched Networks*, Addison-Wesley Longman Publishing Co., Inc., Boston, MA, 1990

G. R. Grimmett and D. R. Stirzaker, *Probability and Random Processes*, Oxford Science Publ, 1998

F. P. Kelly; *Reversibility and Stochastic Networks*, Wiley, London, 1979.

L. Kleinrock, *Queueing Systems, Vol. 1: Theory*, John Wiley, N.Y., 1975.

L. Kleinrock, *Queueing Systems: Vol. 2 Computer Applications*, John Wiley, N.Y., 1976

A. Kumar, D. Manjunath, and J. Kuri: *Communication Networking: An analytical approach*, Morgan-Kaufman (Elsevier), 2004.

J-Y. Le Boudec and P. Thiran; *Network calculus. A theory of deterministic queuing systems for the internet.* Lecture Notes in Computer Science, 2050. Springer-Verlag, Berlin, 2001.

J. R. Norris, *Markov Chains*, Cambridge University Press, 1997.

V. V. Petrov, *Sums of independent random variables*, Translated by A. A. Brown, Springer-Verlag, New York-Heidelberg, 1975.

P. Robert; Stochastic Networks and Queues, Springer, Applications in Mathematics 52, Springer-Verlag, Berlin , 2003.

K. W. Ross: *Multiservice Loss Models for Broadband Telecommunication Networks (Telecommunication Networks and Computer Systems)*, Springer-Verlag, 1995.

R. Serfozo:*Introduction to Stochastic Networks*, Springer, Berlin Heidelberg New York, 2005.

K. Sigman; *Stationary Marked Point Processes: An Intuitive Approach*, Chapman and Hall, NY, 1995.

J. Walrand; *An Introduction to Queueing Networks*, Englewood Cliffs, NJ: Prentice Hall, 1988.

W. Whitt,*Stochastic-process limits,*Springer-Verlag, N.Y., 2002

P. Whittle, *Systems in stochastic equilibrium.* Wiley Series in Probability and Mathematical Statistics: Applied Probability and Statistics. John Wiley & Sons, Ltd., Chichester, 1986.

W. Willinger; *Self-similar Network Traffic and Performance Evaluation*, John Wiley and Sons, N.Y., 2000

R.W. Wolff; *Stochastic Modelling and the Theory of Queues*, Prentice-Hall, N.J., 1989.

JOURNALS AND CONFERENCE PROCEEDINGS

R. R. Bahadur and R. Ranga Rao, *On deviations of the sample mean*, Ann. Math. Statist., 31, 1960, pp. 1015-1027.

F. Baskett, M. K. Chandy, R. R. Muntz, and F. G. Palacios, Open, closed, and mixed networks of queues with different classes of customers, *J. Assoc. Comput. Mach. 22* (1975), 248–260.

G. Bianchi, Performance analysis of the IEEE 802.11 distributed coordination function, Selected Areas in Communications, IEEE Journal on 18 (3), 535-547.

T. Bonald and A. Proutière, Insensitive bandwidth sharing in data networks, *Queueing Syst. Theory Appl.*, 44(1):69–100, 2003.

T. Bonald, A. Proutière, J. Roberts, and J. Virtamo, Computational aspects of balanced fairness, In *In Proceedings of 18th International Teletraffic Congress*, Elsevier Science, 2003, pp. 801–810.

C. Bordenave, D. McDonald, and A. Proutière; A particle system in interaction with a rapidly varying environment: mean field limits and applications. *Netw. Heterog. Media* 5 (2010), no. 1, 31–62.

D. D. Botvich and N. G. Duffield, Large deviations, the shape of the loss curve, and economies of scale in large multiplexers. *Queueing Systems (QUESTA)* 20 (1995), no. 3-4, 293–320.

P. Brémaud, A Swiss Army formula of Palm calculus, *J. Appl. Probab.* 30 (1993), no. 1, 40–51.

P. Brémaud, R. Kannurpatti and R. R. Mazumdar; Event and time averages: A review and some generalizations, *Adv. in Appl. Prob*, 24,(1992), pp. 377-411.

G. L. Choudhury, D. L. Lucantoni, and W. Whitt, *Numerical Transform Inversion to Analyze Teletraffic Models.*,The Fundamental Role of Teletraffic in the Evolution of Telecommunications Networks, Proceedings of the 14th International Teletraffic Congress, J. Labetoulle and J. W. Roberts (eds.), Elsevier, Amsterdam, vol. 1b, 1994, 1043-1052.

S. P. Chung and K. W. Ross, *Reduced load approximation for multi-rate loss networks*, IEEE Trans. on Communications, COM-41 (8), 1993, pp. 1222-1231.

H. Dupuis, B. Hajek, A simple formula for mean multiplexing delay for independent regenerative sources, *Queueing Systems (QUESTA)* 16 (1994), no. 3-4, 195–239.

Z.Dziong, J.W.Roberts (1987), Congestion probabilities in a circuit switched integrated services network, *Performance Evaluation*, Vol.7, N° 3.

A. El Walid and D. Mitra, Effective bandwidth of general Markovian traffic sources and admission control of high speed networks,*IEEE/ACM Transactions on Networking (TON)*, Volume 1 , Issue 3 (June 1993),pp. : 329 - 343.

S. Foss, and T. Konstantopoulos, An overview of some stochastic stability methods, *J. Oper. Res. Soc. Japan 47*, (2004), no. 4, 275–303.

P. Gazdzicki, I. Lambadaris, R.R. Mazumdar, Blocking probabilities for large multi-rate Erlang loss systems,*Adv.Appl.Prob.* 25, 1993 pp. 997-1009.

Guillemin, F. and Mazumdar, R.; Rate conservation laws for multidimensional processes of bounded variation with applications to priority queueing systems, *Methodology and Computing in Applied Probability (MCAP)*, Vol. 6, 2004, pp. 136-159.

C. Graham, Chaoticity on path space for a queueing network with selection of the shortest queue among several. J. Appl. Probab. 37 (2000), no. 1, 198–211.

Graham, Carl Chaoticity for multiclass systems and exchangeability within classes,*J. Appl. Probab.* 45 (2008), no. 4, 1196–1203.

J.P. Haddad, R. R. Mazumdar, and F. J. Piera, Pathwise comparison theorems for stochastic fluid networks, *Queueing Systems (QUESTA)*, Vol. 66, 2010, pp. 155-168.

J-P. Haddad and R. R. Mazumdar, Congestion in large balanced fair systems, *Queueing Systems (QUESTA)*, Special Issue on Network Asymptotics, August 2012. 36 pages. DOI 10.1007/s11134-012-9322-x.

M. T. Hsiao and A. A. Lazar; An extension to Norton's equivalent, *Queueing Systems (QUESTA)*, Vol. 5, 1989, pp. 401-412.

J.Y. Hui, Resource allocation in broadband networks,*IEEE Journal of Selected Areas in Communications*, 1989, 6:1598–1608.

J. S. Kaufman, Blocking in a shared resources environment, *IEEE Trans. Commun.*, COM-29, 10, 1981, pp.1474-1481.

O.Kella, O. and W. Whitt, Stability and structural properties of stochastic storage networks, *J. Appl. Probab.*, Vol 33 (4), 1996, pp. 1169-1180.

F.P. Kelly, Blocking probabilities in large circuit-switched networks,*Adv.Appl.Prob.* **18**, 1996, pp.473-505.

F. P. Kelly, Loss networks, *Annals of Applied Probability*, 1 (1991), 319-378.

F.P. Kelly, Effective Bandwidths at Multiclass Queues, *Queueing Systems (QUESTA)*, 9 (1991), pp. 5-16.

F.P. Kelly, Charging and rate control for elastic traffic . *European Transactions on Telecommunications*, volume 8 (1997) pages 33-39

G. Kesidis, J. Walrand, and C.S. Chang, Effective bandwidths for multiclass Markov fluids and other ATM sources, *IEEE/ACM Transactions on Networking (TON)*, 1993, Vol 1, 424-428.

T. Konstantopoulos, and G. Last, On the dynamics and performance of stochastic fluid systems, *J. Appl. Probab.*, 37 (2000), no. 3, 652–667.

N.B. Likhanov, and R. R. Mazumdar, Cell loss asymptotics for buffers fed with a large number of independent stationary sources. *J. Appl. Probab.* 36 (1999), no. 1, 86–96.

N. B. Likhanov, R. R. Mazumdar, and F. Theberge, Providing QoS in Large Networks: Statistical Multiplexing and Admission Control, in *Analysis, Control and Optimization of Complex Dynamic Systems*, E. Boukas and R. Malhame , eds., Springer 2005, pp 137-1169.

G. Louth, M. Mitzenmacher and F.P. Kelly, Computational complexity of loss networks, *Theoretical Computer Science*, Vol. 125 (1), 1994, pp. 45-59.

R. M. Loynes, The stability of a queue with non-independent inter-arrival and service times, *Proc. Cambridge Philos. Soc.* 58 1962 497–520.

L. Massoulié, Structural properties of proportional fairness: Stability and insensitivity, *Ann. Appl. Probab.*, 17(3):809–839, 2007.

R.R. Mazumdar, R. Kannurpatti, and C. Rosenberg.; On rate conservation for non-stationary processes, *J. of Appl. Probab.*, Vol. 28, No. 4, Dec. 1991, pp. 762-770.

R. R. Mazumdar, R., Badrinath, V., Guillemin, F. and Rosenberg, C.; A note on the pathwise version of Little's formula, *Operations Research Letters*, Vol. 14, No. 1, 1993, pp. 19-24.

R. R. Mazumdar, V. Badrinath, F. Guillemin, and C. Rosenberg, On Pathwise Rate Conservation for a Class of Semi-martingales, *Stoc. Proc. Appl*, 47, 1993, pp. 119-130.

B. Melamed, Characterizations of Poisson traffic streams in Jackson queueing networks, *Adv. in Appl. Probab.* 11 (1979), no. 2, 422–438.

B. Melamed, On the reversibility of queueing networks. *Stochastic Process. Appl.* 13 (1982), no. 2, 227–234.

D.Mitra, J.Morrison, Erlang capacity and uniform approximations for shared unbuffered resources, *IEEE/ACM Transactions on Networking*, vol.2, N.6, 1994, pp.558-570.

M. Mitzenmacher, A. W. Richa, and R. Sitaraman; The power of two random choices: a survey of techniques and results. *Handbook of randomized computing, Vol. I, II,* 255–312, Comb. Optim., 9, Kluwer Acad. Publ., Dordrecht, 2001.

M. Miyazawa, A Formal Approach to Queueing Processes in the Steady State and their Applications, *J. of Applied Probability*, 16, 1979, pp. 332-346.

M. Miyazawa, Rate Conservation Laws: A Survey, *Queueing Systems Theory (QUESTA)*, 15, 1994, pp.1-58.

J. Mo, and J. Walrand, Fair End-to-End Window-Based Congestion Control. *IEEE/ACM transactions on Networking*, 2000(8): 556-567.

P. Nain, Qualitative properties of the Erlang blocking model with heterogeneous user requirements , *Queueing Systems (QUESTA)*, Vol. 6 (1), 1990, pp. 189-206.

C. Rainer, and R. R. Mazumdar ; A note on the conservation law for continuous reflected processes and its application to queues with fluid inputs, *Queueing Systems (QUESTA)*, Vol. 28, Nos. 1-3,(1998), pp.283-291.

S. Ramasubramanian, A subsidy-surplus model and the Skorokhod problem in an orthant, *Math. Oper. Res.*, 25 (3), 2000, pp. 509-538.

S. Resnick, and G. Samorodnitsky; Limits of on/off hierarchical product models for data transmission, *Ann. Appl. Probab.*, 13 (2003), no. 4, 1355–1398.

J. W. Roberts, A service system with heterogeneous user requirements: Application to multi-service telecommunications systems, *Perf. of Data Comm. Syst. and their Applications*, G. Pujolle ed., pp.423-431, North-Holland, 1981.

J. W. Roberts and L. Massoulié, Bandwidth sharing and admission control for elastic traffic. *Telecommunication Systems*, 15:185–201, 2000.

R. Schassberger, The insensitivity of stationary probabilities in networks of queues. *Adv. in Appl. Probab.* 10 (1978), no. 4, 906–912. Correction:*Adv. in Appl. Probab.* 10 (1978), no. 4, 906–912.

A.D. Skorokhod, Stochastic equations for diffusions in a bounded region, *Theory of Prob. and its Appl.*, Vol 6, 1961, pp. 264-274.

A. Simonian, F. Théberge, J. Roberts, and R. R. Mazumdar, ; Asymptotic estimates for blocking probabilities in a large multi-rate loss network, *Advances in Applied Probability*, Vol. 29, No. 3, 1997, pp. 806-829.

F. Theberge and R.R. Mazumdar(1996), A new reduced load heuristic for computing blocking in large multirate loss networks, *Proc. of the IEE- Communications*, Vol 143 (4), 1996, pp. 206-211.

N.D. Vvedenskaya, R.L. Dobrushin, and F. I Karpelevich, A queueing system with a choice of the shorter of two queues—an asymptotic approach. *(Russian) Problemy Peredachi Informatsii 32* (1996), no. 1, 20–34; translation in *Problems Inform. Transmission 32* (1996), no. 1, 15–27.

N. Walton, Insensitive, maximum stable allocations converge to proportional fairness, *Queueing Systems*, 68:51–60, 2011.

P. Whittle, Partial balance and insensitivity. *J. Appl. Probab.* 22 (1985), no. 1, 168–176.

P. Whittle, Partial balance, insensitivity and weak coupling,*Adv. in Appl. Probab.* 18 (1986), no. 3, 706–723.

W. Willinger, W.E. Leland, M.S. Taqqu, and D.V. Wilson, On the Self-Similar Nature of Ethernet traffic,*IEEE/ACM Transactions on Networking*, Vol. 2 (1), 1994, pp. 1-15.

H. Yaiche, R. R. Mazumdar, and C. P. Rosenberg, A game theoretic framework for bandwidth allocation and pricing in broadband networks, *IEEE/ACM Trans. on Networking*, 8(5):667–678, 2000.

S. Zachary and I. Zeidins, Loss networks. In *Queueing networks*, R. J. Boucherie and N.M. van Dijk eds, Internat. Ser. Oper. Res. Management Sci., Vol.154, Springer, New York, 2011.

Author's Biography

RAVI R. MAZUMDAR

Ravi R. Mazumdar was born in 1955 in Bangalore, India. He was educated at the Indian Institute of Technology, Bombay (B.Tech. 1977), the Imperial College of Science, Technology and Medicine, London, UK (MSc, DIC 1978), and received his Ph.D. from the University of California, Los Angeles (UCLA) in 1983.

He was an Assistant Professor of Electrical Engineering at Columbia University, NY (1985-88); Associate and Full Professor at the Institut National de la Recherche Scientifique, Montreal, Canada (1988-97). From 1996- 2000 he held the Chair in Operational Research at the University of Essex, Colchester, UK , and from 1999-2005 was Professor of ECE at Purdue University, West Lafayette, USA. Since Fall 2004 he has been with the Department of Electrical and Computer Engineering at the University of Waterloo, Canada as a University Research Chair Professor. Since 2012 he is also a J.D. Gandhi Distinguished Visiting Professor at the Indian Institute of Technology, Bombay.

Dr. Mazumdar is a Fellow of the IEEE and the Royal Statistical Society. He was a recipient of the IEEE INFOCOM Best Paper Award in 2006 and was Runner-up for the Best Paper at the INFOCOM in 1998.

His research interests are in stochastic modeling and analysis with applications to wireline and wireless networks; in game theory and its applications to networking; and in stochastic analysis and queueing theory.

Index